46 Science Fair Projects for the Evil Genius

Evil Genius Series

Bike, Scooter, and Chopper Projects for the Evil Genius

Bionics for the Evil Genius: 25 Build-it-Yourself Projects

Electronic Circuits for the Evil Genius: 57 Lessons with Projects

Electronic Gadgets for the Evil Genius: 28 Build-it-Yourself Projects

Electronic Games for the Evil Genius

Electronic Sensors for the Evil Genius: 54 Electrifying Projects

50 Awesome Auto Projects for the Evil Genius

50 Model Rocket Projects for the Evil Genius

51 High Tech Practical Jokes for the Evil Genius

46 Science Fair Projects for the Evil Genius

Fuel Cell Projects for the Evil Genius

Mechatronics for the Evil Genius: 25 Build-It-Yourself Projects

MORE Electronic Gadgets for the Evil Genius: 40 NEW Build-It-Yourself Projects

101 Outer Space Projects for the Evil Genius

101 Spy Gadgets for the Evil Genius

123 PIC® Microcontroller Experiments for the Evil Genius

123 Robotics Experiments for the Evil Genius

PC Mods for the Evil Genius: 25 Custom Builds to Turbocharge Your Computer

Programming Video Games for the Evil Genius

Solar Energy Projects for the Evil Genius

22 Radio and Receiver Projects for the Evil Genius

25 Home Automation Projects for the Evil Genius

46 Science Fair Projects for the Evil Genius

BOB BONNET
DAN KEEN

New York Chicago San Francisco Lisbon
London Madrid Mexico City Milan New Delhi
San Juan Seoul Singapore Sydney Toronto

Library of Congress Cataloging-in-Publication Data

Bonnet, Robert L.
  46 science fair projects for the evil genius / [Bob Bonnet, Dan Keen].
      p. cm. — (Evil genius series)
  ISBN 978-0-07-160027-9 (alk. paper)
1. Science projects. I. Keen, Dan. II. Title. III. Title: Forty-six science
fair projects for the evil genius.
  Q182.3.B669 2008
  507.8—dc22   2008008078

McGraw-Hill books are available at special quantity discounts to use as premiums and sales promotions, or for use in corporate training programs. To contact a representative please visit the Contact Us pages at www.mhprofessional.com.

46 Science Fair Projects for the Evil Genius

1 2 3 4 5 6 7 8 9 0   QPD/QPD   0 1 3 2 1 0 9 8

ISBN 978-0-07-160027-9
MHID 0-07-160027-2

Sponsoring Editor
Judy Bass

Acquisitions Coordinator
Rebecca Behrens

Editing Supervisor
David E. Fogarty

Project Manager
Patricia Wallenburg

Copy Editor
Marcia Baker

Proofreader
Paul Tyler

Indexer
Karin Arrigoni

Production Supervisor
Richard A. Ruzycka

Composition
Patricia Wallenburg

Art Director, Cover
Jeff Weeks

About the Authors

Bob Bonnet, who holds a master's degree in environmental education, has been teaching science for over 25 years. He was a state naturalist at Belleplain State Forest in New Jersey. Mr. Bonnet has organized and judged many science fairs at both the local and regional levels. He has served as the chairman of the science curriculum committee for the Dennis Township School system, and he is a Science Teaching Fellow at Rowan University in New Jersey.
Mr. Bonnet is listed in "Who's Who Among America's Teachers."

Dan Keen holds an Associate in Science degree, majoring in electronic technology. Mr. Keen is the editor and publisher of a county newspaper in southern New Jersey.

He was employed in the field of electronics for 23 years, and his work included electronic servicing, as well as computer consulting and programming. Mr. Keen has written numerous articles for many computer magazines and trade journals since 1979. He is also the coauthor of several computer programming books. For ten years, he taught computer courses in community education programs in four schools. In 1986 and 1987, Mr. Keen taught computer science at Stockton State College in New Jersey.

Together, Mr. Bonnet and Mr. Keen have had many articles and books published on a variety of science topics for international publishers, including McGraw-Hill.

Contents

Contents

Introduction

Welcome to the exciting exploration of the world around us. . . the world of science! Researching a project for entry into a science fair gives us a glimpse into the marvels of this world.

Participating in a science fair is not only enjoyable, but it also encourages logical thinking, involves doing interesting research, develops objective observations, and gives experience in problem solving.

Before you do any project, discuss it in detail with a parent or science instructor. Be sure they understand and are familiar with your project.

Science fair projects must follow a procedure called the scientific method. This procedure is also used by actual scientists. First, a problem or purpose is defined. A hypothesis or prediction of the outcome is then stated. Next, a procedure is developed for determining whether or not the hypothesis was correct. Do not think that your science project is a failure if the hypothesis is proven to be wrong. The idea of the science fair project is either to prove or disprove the hypothesis. Learning takes place even when the results are not what you expected. Thomas Edison tried over a thousand different materials before he found one that would work best in his light bulb. Edison said he failed his way to success!

Generally, school science fairs have 12 standard categories under which students can enter their projects: behavioral and social, biochemistry, botany, chemistry, Earth and space, engineering, environmental, physics, zoology, math and computers, microbiology, and medicine and health.

Some projects may involve more than one science discipline. A project that involves using different colors of light to grow plants could fall under the category of either botany or physics. This crossing over of sciences may allow you to choose between two categories in which to enter your project. It can give you an edge at winning a science fair by entering your project in a category where there are fewer competitors or avoiding a category where other entries are of particularly outstanding quality.

In this book, we present a wide variety of project ideas for all 12 science fair categories. Select a topic you find interesting, one you would like to research. This will make your science fair experience a very enjoyable one. Many projects in this book are merely "starters," which you can expand on and then create additional hypotheses for.

Know the rules of your school's science fair before you decide on a project topic. Projects must follow ethical rules. A project cannot be inhumane to animals. Never interfere with ecological systems. Use common sense.

Safety

When planning your science fair project, safety must be your first consideration. Even seemingly harmless objects can become a hazard under certain circumstances. Know what potential hazards you are faced with before you start a project. Take no unnecessary risks. Have an adult or a science instructor present during all phases of your project. Be prepared to handle a problem even though none is expected (for example, keep heat gloves or oven mitts handy when you work around a hot stove). Wear safety glasses when appropriate.

Be Especially Aware of These Hazards

- Sharp objects: Construction tools (hammer, saw, knife, scissors, drill). Be careful how you pick up sharp tools and glass objects, which can fragment and become sharp objects.

- Fire: Cooking fat can catch on fire; alcohol has a low flash point. To boil alcohol, use a "double boiler." First, bring a pot of water to a boil. Next, turn off the stove burner. And then, lower a test tube filled with alcohol into the water.

- Chemicals: Keep everything out of the reach of children that specifies "keep out of the reach of children" on the label (alcohol, iodine, and so forth). Know

what materials you are working with that have extreme pH levels (acids, bases).

- Allergens: When growing mold in sealable plastic bags, keep the bags closed during and after the project. When the project is over, discard the plastic bags without ever opening them, so mold is contained and does not become airborne.

- Carcinogens, mutagens: Stand away from microwave ovens when in use.

- Water and electricity don't mix. Use caution whenever both water and electricity are present (as with a fish tank heater that must be plugged into a wall outlet). Use only UL-approved electrical devices.

- Heat: Use heat gloves or oven mitts when you deal with hot objects. When using a heat lamp, keep away from curtains and other flammable objects. Be aware that glass may be hot, but it might not give the appearance of being hot.

- Secure loose clothing, sleeves, and hair.

- Wash your hands. When you return home after touching surfaces at public places, be sure to wash your hands to avoid bringing bacteria into your home.

- Rivers, lakes, oceans: Do not work near or around large bodies of water without an adult present, even if you know how to swim.

- Nothing should be tested by tasting it.

- Be aware of others nearby. A chemical reaction, for example, could cause a glass container to shatter or a caustic material to be ejected from a container. Keep others in the room at a safe distance or have them wear proper safety protection.

- Thermometers made of glass have the potential to break and cause glass to shatter.

- Be aware of gas products that may be created when certain chemicals react. Such projects must be carried out in a well-ventilated area.

- Never look directly at the Sun. Do not use direct sunlight as a source of light for microscopes.

- Loud sounds can be harmful to your hearing.

Being aware of these possible hazards and working with adult supervision should ensure a safe and enjoyable project experience.

What Makes a Good Science Fair Project?

A good science fair project is either something that is unique or it is something that is already common, but done uniquely. For example, many elementary students construct a small model of a volcano, and then use the reaction of vinegar and baking soda to make it "erupt." Such a project could have a unique "twist" to it by hypothesizing that some other substance or chemical reaction would effervesce and give a better eruption.

A good project is also one where the student has done a solid background study and fully understands the project. It's fine to have an adult or even a science professional assist a student in their project, but a judge will expect the student to understand the project and be able to articulate the work to the judges and others attending a science fair. A project will be judged on its completeness. Students should look at their projects as if they are the judges and check for any deficiencies. Presentation is important, but many science fairs weigh more heavily on the science aspect of projects.

Good luck with your project!

Water, Water, Everywhere

The effect of fresh water and coastal salt water flooding on lawns

Suggested Entry Categories

- Biochemistry
- Botany
- Chemistry
- Earth Science
- Environmental Science

Overview

People often pay a high price to purchase land and build a house along the coast, or along a scenic river or stream. The view is always magnificent; the fresh air and walking along the shore are especially healthy. However, not only is the initial cost of real estate expensive, but so is property upkeep. For coastal homes, the salt air and strong winds act as sand blasters to pit the metal on door knobs, window casings, and house paint. Coastal storms are an ever-present threat, too. Another risk for home owners living along rivers or oceans is flooding.

Even a small flood can damage the beautiful and expensive lawns around a home.

Is more damage done to a lawn by fresh water river flooding or coastal salt water flooding?

Hypothesis

Hypothesize that more damage to lawns is caused by coastal salt water flooding than by the flooding of a fresh water stream or river.

Materials' List

- Two large dishpans
- Several pieces of 1×2 lumber
- Small nails
- Use of a hammer and hand saw
- Several feet of cheesecloth
- Instant synthetic sea salt mix (available inexpensively from school science supply catalogs)
- Water
- Grass seed
- Potting soil
- Staple gun
- Funnel
- Scissors
- Kitchen measuring cup
- Four empty plastic gallon milk or water jugs
- A warm, lighted area indoors, but not in direct sunlight
- Several weeks of time, because we are dealing with germination and growth

Procedure

Grass seed will germinate and grow in two wooden frames of potting soil. Both

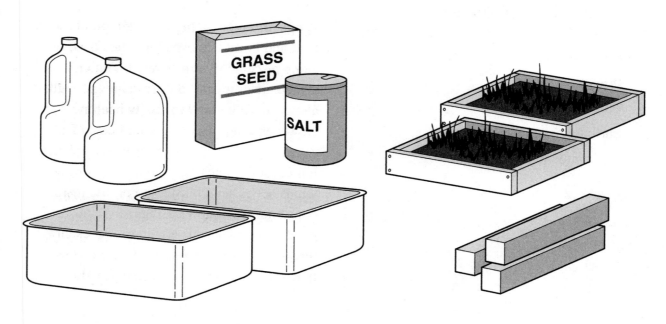

"miniature lawns" will be kept next to each other to maintain the same environment, each receiving an equal amount of light and being kept at the same temperature.

The variable in this project is the exposure of one lawn to severe salt water flooding, and the other to fresh water flooding.

Locate two large rectangular dishpans, used for washing dishes.

With several pieces of 1×2 wood and small nails (or screws), construct two rectangular frames that fit inside the dishpans. Cut a rectangular piece of cheesecloth to cover a frame. Staple the cheesecloth to the wooden frame, keeping it pulled tight. Repeat for the other frame. Now, turn the frames upside down and fill them with potting soil. The cheesecloth holds the potting soil in the frames, but it allows excess water to pass through.

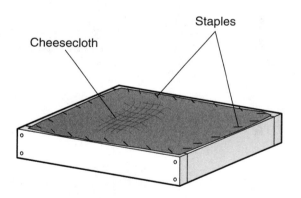

Place the two dishpans in a warm, well-lit area, but not in direct sunlight. Across the top of each dishpan, lay two pieces of wood, and set a wooden frame over each one. The pieces of wood will support the frames over the dishpans. Pour some grass seed in a kitchen measuring cup, and then spread the seeds out on the soil of one of the frames.

Pour an equal amount of seed into the cup, and spread over the soil in the second frame. Lightly cover the seeds with soil and moisten the soil in the frames.

Make observations daily and keep the soil moist (but not soaked), watching for germination. Equal amounts of water should be given to each lawn frame. Allow the grass to grow until the blades are around one to two inches tall. When that happens, continue to the next step.

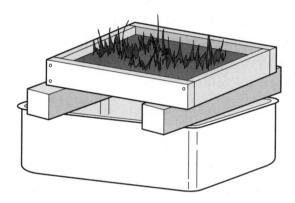

Fill four 1-gallon plastic milk or water jugs with tap water. To two of the jugs, add a synthetic sea salt mix, as per the instructions on the package. These mixes are available at science shops and through science catalogs from your school science teacher. They are inexpensive. The mix contains all the essential major and minor elements to create a solution that closely matches ocean water.

Remove the two wooden supports on one flat and lower it into the dishpan. Slowly, so you don't cause erosion of the soil, pour the two gallons of salt water solution into the dishpan. Leave the water in the pan for one hour, and then pour it off. You can save the solution by using a funnel and pouring it back into the bottles. Lift the frame out of the dishpan and place the wood supports back under it, so the soil can drain.

Similarly, lower the other lawn frame into its dishpan and flood it with two gallons of fresh water. Let it sit for one hour, and then pour off the water and place the supports back under the frame.

Allow the lawn frames to dry for two days. Make observations, looking for any changes in grass (color, turgor, and so forth) Record your observations. If no differences are observed, repeat the flooding procedure on the third day. Then, again allow to dry for three days. Continue to repeat the flooding and drying process until you see an observable difference.

Results

Write down the results of your experiment. Document all observations and data collected.

Conclusion

Come to a conclusion as to whether or not your hypothesis was correct.

Something More

1. If a lawn is killed by salt water flooding, can the home owner simply replant grass seed on the lawn once the flooding has passed, or is the soil made unfit for growing new plants? If the soil is unfit, how can it be cleared of salt and made ready to support life again? Should a home owner turn on his lawn sprinklers after a flood to dilute and wash the salts and other materials left by the sea water?

2. Is one type of seed more tolerant of salt water flooding? This would be important to know for landscapers and home owners in seashore communities.

3. Does pouring salt in the cracks in a sidewalk or driveway kill any grass or weeds that grow there? If so, this would be a safe way to kill unwanted weeds, because salt is not a hazard to people or pets.

Project 2

Who's Home?

Determining whether or not organisms other than birds live in birds' nests

Suggested Entry Categories

- Environmental Science
- Microbiology
- Zoology

Purpose or Problem

The purpose is to determine if a bird's nest is home to more organisms than just birds.

Overview

The Earth is teeming with life. Just think how many things are alive within 100 feet of where you are right now: worms in the ground, flowers, trees, grasses, an insect on a window screen, a microscopic mite on your pillow, mold on a piece of bread left uncovered in the kitchen, perhaps even a family member in the next room. You may hear the peaceful singing of a bird building a nest outside your window.

Birds lack the carpentry skills of humans, and they obviously don't have the use of arms or hands. Yet, they are quite capable of constructing nests that are structurally sufficient for the laying of eggs and raising their young.

Nature provides all the nest-building materials a bird needs: twigs, feathers, animal hair, straw, moss, leaves, pebbles, blades of grass, and even some items provided by humans—a piece of yarn, string, or paper.

Because nest building materials come from nature, and life is abundant all around us, do you think other things are living in birds' nests besides birds?

Hypothesis

Hypothesize that you can find other forms of life besides birds in a bird nest.

Materials' List

- Bird nest containing baby birds
- Desk lamp that uses a standard 60 to 75 watt incandescent bulb
- Large funnel
- Clear jar about the size of a drinking glass
- High-power hand lens (magnifying glass)
- Microscope
- Small plastic bag
- Ten petri dishes with agar

Procedure

Scout around the trees on your property or in your neighborhood and look for a bird's nest with baby birds inside. The nest must be within reach or able to be easily and safely retrieved (you don't want one that is 50 feet in a tree top).

Once you locate a suitable nest, watch it once or twice a day, waiting for the day when the last baby bird leaves the nest. Do not get too close or disturb the nest in any way.

As soon as possible after you see all the birds are gone and the nest is no longer used by the mother bird, carefully remove the nest and place it in a plastic bag.

Take the nest home (or to school), but do not take it inside your house, just in case it contains insects or microscopic life that would not be good to have inside your home. Set the nest on a picnic table, a portable card table, or on a workbench in a garage. To collect tiny insects that may be living in the nest, place a large-mouth funnel in a clear jar. Then, set the nest in the mouth of the funnel. Position a desk lamp over the top of the nest, but keep a space of several inches between the lamp's bulb and the nest to prevent the nest from getting hot. The incandescent bulb in the desk lamp should be about 60 or 75 watts. The heat from the bulb may drive any insects down into the glass, as they try to escape the heat. Leave the bulb on for one hour, and then carefully examine the glass for anything that has been collected. During the time the light is on, do not leave it unattended. Watch that the nest is not becoming too hot (to avoid a fire hazard and

Results

Write down the results of your experiment. Document all observations and data collected.

Conclusion

Come to a conclusion as to whether or not your hypothesis was correct.

harming anything that may be living in the nest). Use a high-power magnifying glass to examine any material that falls into the jar. Attempt to identify the organisms using field guides and other reference materials.

Next, check for the presence of smaller organisms in the nest. Do this by taking ten pieces from different locations on the nest and wiping them several times on agar in petri dishes. Cover the petri dishes and place them in a warm, dark location. After two weeks, examine each petri dish under a microscope. Never open any of the petri dishes once they have been closed. Eventually, when the project is over, dispose of the petri dishes, continuing to keep them sealed shut.

Something More

1. Can you locate other similar nests in your area that would indicate they were built by the same species of bird? The mother bird, the structure of the nest, and the size and designs on the egg shells will help you identify the species of bird using the nest. A good book on birds will be necessary to help you identify the species. Then, run the same tests as you did previously. Are the same organisms found in these nests?

2. What else did you find in the nest: leftover food, a piece of egg shell?

3. What is the composition of the nest? Can you identify other materials used making the nest?

4. How are nests adapted for rain? How are they adapted to ward off attacks from other animals?

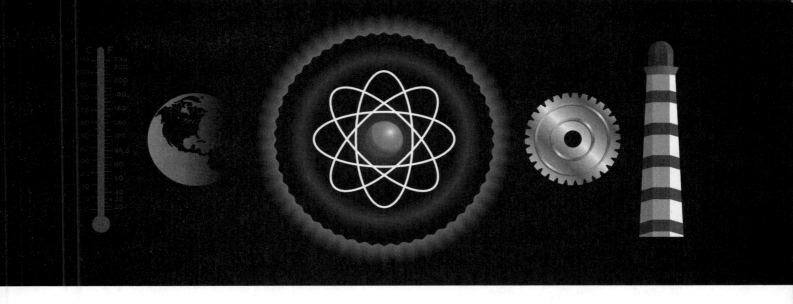

Go with the Flow

Lighthouses are cylindrically shaped, so they can structurally withstand high-velocity winds

Suggested Entry Categories

- Earth Science
- Engineering
- Environmental Science
- Physics

Purpose or Problem

Lighthouses must be built along the coast and they must be tall, but that subjects these structures to fierce winds. Builders have learned to make the shape of lighthouses round, causing air to flow around them with less resistance, and allowing them to withstand strong winds.

Overview

Sea coasts are beautiful, but it's not unusual for them to experience violent storms with furious winds. Through the years, builders have had to take this environment into account when they design lighthouses. These unique buildings that have aided storm-driven sailors for centuries must be constructed to withstand hard winds and weather. Lighthouses are also used for identification by those at sea to help them get their bearings as to where they are in relation to the coast, a shoal, or a safe harbor.

A good defense against the wind is to offer as little resistance as possible and to deflect the moving air past the structure, so it flows smoothly around it. Have you ever held a large sheet of plywood and tried walking with it on a windy day? Think about a sail on a sailboat; it presents a lot of resistance to the wind and uses the wind's force to propel the boat.

A building with the shape of a cylinder guides the air flow around it and allows the air to continue behind it. Such a structure can withstand higher winds, as it has less force than on a similar structure that catches the wind. Therefore, you may have noticed from seeing pictures or visiting lighthouses that most of them are cylindrical in shape. Now you know why!

Hypothesis

Hypothesize that moving air flows more easily around a cylindrically shaped object

than one with a flat surface facing the wind and, therefore, offers less resistance to wind.

Materials' List

- Thirty-three (33) long straight pins
- Spool of thread
- Piece of plywood 1 foot square
- Piece of balsa wood 1 foot square (or several smaller pieces that can be laid side by side to cover a 1-foot-square area)
- Glue
- Ruler
- Pencil
- A cylindrically shaped object between 3 and 3½ inches in diameter (a glass jar or a can of fruit—we recommend a cardboard container for bread crumbs)
- Two pieces of 2×4 lumber, each about 5 or 6 inches long
- Hair dryer
- Pair of scissors
- Possible adult supervision needed

Procedure

The *constant* in this project is the velocity of the approaching air. The *variable* is the shape of the object around which the air must flow.

For us to see the pattern of air flow around an object, we must first construct a device that visually shows us the presence and direction of air flow (an "air flow table"). Obtain a piece of plywood that is at least 12 inches square. Glue a 12"×12" sheet of balsa wood on top, or attach it by using several very small screws or nails. If you cannot get a single sheet of balsa wood that big, use several smaller pieces, lay them side by side, and carefully cut them with a utility knife to form a 1-foot-square area. Use extreme caution when you work with a utility knife.

Using a ruler, mark a grid pattern of lines at 1½ inch increments, both horizontally and vertically, on the balsa wood. At the point where each line intersects, carefully push a long straight pin into the balsa wood with your thumb. As shown in the illustration, do not put pins near the front of the board in the locations covered by the shaded circle. This is where the objects under test will be placed.

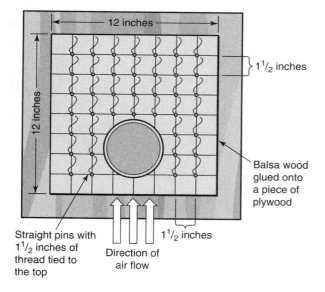

12 inches

1½ inches

12 inches

Balsa wood glued onto a piece of plywood

Straight pins with 1½ inches of thread tied to the top

Direction of air flow

1½ inches

Tie thread onto each pin, and position it near the pin's head. Using scissors, trim the thread to a length of 1½ inches. You can use a small drop of glue on the pin to hold the thread securely in place. This is helpful if you plan to move the project from home to a classroom or a science fair.

Secure two pieces of 2×4 wood together, each about 5 inches tall (use glue, string, screws, or nails). This makes a structure that is almost square on four sides. Stand it upright in the empty space on the balsa board.

Hold a hair dryer in front of the balsa board and aim it directly at the 2×4 wood block. Place the hair dryer on a setting that blows the most air. If the hair dryer has a cool setting, use it, because heat is not needed. Observe the pattern of the threads. Do the threads directly behind the block move?

Remove the wood block and replace it with a cylindrically shaped object that is about 3½ inches in diameter. A large cardboard bread-crumb container works well.

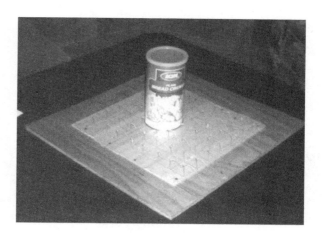

With the hair dryer in the same position and at the same setting, do the directions of the threads look different? Do the threads directly behind the cylinder now indicate a strong air flow?

Results

Write down the results of your experiment. Document all observations and data collected.

Conclusion

Come to a conclusion as to whether or not your hypothesis was correct.

Something More

1. Experiment with structures of different shapes (try a pyramid, for example). Observe the flow of air around them.

2. Construct a *stream table*, a device through which water can flow and objects can be inserted to study how shape affects the flow of water. Food coloring dye is dropped into that water to better visualize the pattern of water flow. Prove that the behavior of the flow of water and the behavior of the flow of air around an object are similar because they are both "fluids."

Kinetic Pendulum

Examining the relationship between the arc distance a pendulum travels and the swing period time

Suggested Entry Categories

- Math & Computers
- Physics

Purpose or Problem

The purpose is to understand one of the principles of pendular motion.

Overview

A *pendulum* is a weight hung by a tether (a rope, string, or rod) from a fixed point, and made to swing. When the pendulum is pulled away from its motionless hanging state (perpendicular to the Earth), the weight gains potential or stored energy. When released, the *potential* energy is turned into *kinetic* or working energy.

Once released, the pendulum is pulled down toward the Earth by gravity, but it does not stop when it returns to the Earth's perpendicular plane (called *plumb*). At that point, the moving pendulum has *momentum* (mass multiplied by velocity), which causes it to continue to swing past the plumb point, until the force of gravity slows it to a stop. The pendulum then swings back through the plumb and up to the point where it was first released. This swing out and back is called one *oscillation period*. Then, once again, gravity continues its effect, and the pendulum continues to swing back and forth.

If it were not for the friction with air against the pendulum and the friction at the point where it is secured to a fixed point, the swinging would continue indefinitely.

Many early scientists, including Lord Kelvin (1824–1907), Jean Foucault (1819–1868), and Galileo (1564–1642), devoted time to the study of the natural laws of pendular motion. Galileo was reported to note, while sitting in church, that a chandelier swung with the same time period, regardless of whether it was swinging in a small arc or a large arc (the sermon must not have been very interesting that day!). This project will attempt to prove this natural law of pendular motion discovered by Galileo.

Hypothesis

Hypothesize that the swing period of a pendulum with a fixed rope length is the same, regardless of the arc distance traveled. (Because of air resistance and other factors, we will state this hypothesis is true for the first five oscillation periods of our constructed apparatus.)

Materials' List

- Two bowling balls of the same weight
- Two plastic bags with handles (used at grocery and retail stores)
- Rope
- String
- Child's outdoor swing set
- Yard stick or tape measure
- Large, heavy metal washer
- A day with negligible or no wind
- A friend to assist
- Possible adult supervision needed

Procedure

The mass of the bowling balls, the tape measure, the length of the ropes, and the environment (temperature, humidity, wind) are held constant. The distance each ball is pulled back from the plumb line is varied.

Find two bowling balls of equal weight and set each one in a plastic bag, the kind used at grocery and retail stores to carry products home. Although these bags are very thin, they are strong and have convenient handles. Bowling balls can be hazardous if they fall on your feet. **Pay extra attention and take safety precautions when you work with the bowling balls.** Place them on the ground, never on a table where they could unexpectedly roll off.

Tie a long piece of rope through the two handles on one of the bags. Tie another long

piece of rope through the handles on the other bag.

Tie a long piece of string onto a heavy metal washer. From a child's backyard swing set, tie the other end of the string to the top pipe, letting the washer hang about one or two inches from the ground. Be sure the washer hangs freely and does not touch any of the swings.

Similarly, tie the two bowling balls in their bags from the top pipe. Be sure they hang freely and do not touch any of the swings or each other. Using a tape measure, make the distance from the top pipe to the top of each bowling ball exactly the same length.

The washer on a string acts as a *plumb line*, also called a *plumb bob*, which is a

weighted line that is perpendicular to the ground.

Pull one of the bowling balls back about four feet from the plumb line. Have your friend pull the other ball back about one foot. On the count of three, both of you should let go of the balls at the same time. It is important for both of you to let go simultaneously.

Notice that even though your ball has farther to travel, it will cross over the plumb-line point at the same time as the ball your friend let go.

Watch the balls swing through five periods, and note they are still hitting the plumb line at the same time, proving the hypothesis correct.

Because of other variables, including friction with the air (one ball moves through more air than the other and, thus, experiences more friction), the balls may eventually stop meeting at the plumb point.

You may want to measure the distance the bowling balls travel by measuring the length of the arcs. When the ball is pulled back one foot from plumb, how many degrees is the angle from plumb? How many degrees is the angle when the ball is pulled back four feet?

Results

Write down the results of your experiment. Document all observations and data collected.

Conclusion

Come to a conclusion as to whether or not your hypothesis was correct.

Something More

1. A common natural law of gravity and astronomy (celestial mechanics) that also applies to pendulums is the *inverse-square law*, which states the following: if one pendulum is twice as long as another, the longer one will have a period that is "one over the square of two," or one fourth, as fast:

$$\frac{1}{4}$$

   Prove this expression by experimentation.

2. Pendular mechanisms have been used throughout history to keep time. Construct a pendulum that completes one period in one second (clue: the length of the string should be about 39.1 inches).

3. Research the work of the English scientist Lord Kelvin and his discoveries with bifilar pendulums (having two strings instead of one).

4. Research the work of the French scientist Foucault, who used a large iron ball on a wire to show that the Earth rotates.

5. Could you use pendulums or plumb lines to detect earthquakes or other vibrations in the Earth?

Melody Camouflage

Erroneously perceived sound while masked by noise

Suggested Entry Category

- Behavioral & Social

Purpose or Problem

The purpose is to prove that often people "hear" what they expect to hear, even if the sound is not present.

Overview

Have you ever listened to a blank cassette tape on a stereo that had the volume set loud? All you hear is a high-pitched hissing sound. This "noise" is due to the nature of tape as a recording medium.

"Noise" in the reproduction of audio is unwanted sound caused by the tape and electronic components in the amplifier. This hissing sound was not part of the original source material.

Tape hiss has plagued the music and audio industry for years. Today, electronics has advanced to the point that hiss caused by electronic circuitry is almost nonexistent, especially on professional audio equipment. Another technological breakthrough, the compact disc (CD), has made a tremendous advancement in reducing audible hiss in recorded music.

A psychoacoustical masking effect takes place when music is played at high volumes. Noise such as tape hiss seems to disappear during loud passages of music.

Another interesting behavioral effect is that we sometimes hear what we expect to hear. In this project, we record music and "white noise" together, and then gradually reduce the music until only the white noise remains. Will people claim to continue to "hear" the music in the presence of white noise, even after it is turned off?

Hypothesis

Hypothesize that, when tested, a greater number of your friends and classmates will continue to "hear" music even after the music has completely stopped, while the presence of a high level of white noise remains.

Materials' List

- Stereo audio mixer
- Blank cassette tape
- Headphones
- Cassette player
- Cassette recorder
- Cassette tape of a popular song all your test subjects are very familiar with

- Electronic music synthesizer keyboard with a white noise sound
- 20 friends and classmates
- Stop watch or a clock/watch with a seconds display

Procedure

The volume level of the white noise will be held constant. The volume level of the music will be varied.

You need to make a cassette tape with which to test your subjects. The tape must contain white noise recorded at a high volume, along with a song your test subjects are very familiar with.

To do this, you need a source of white noise, such as a musical instrument synthesizer keyboard, which has a white noise–like setting. Connect the synthesizer's output into an audio mixer. Into another channel of the mixer, connect the output of a cassette tape player. The output of the mixer must then feed another cassette recorder that has a blank tape to record the results.

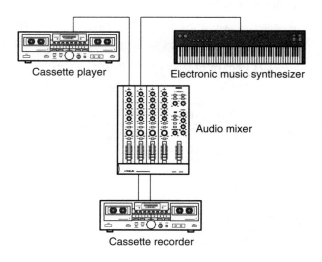

Cassette player   Electronic music synthesizer

Audio mixer

Cassette recorder

If you do not have access to an audio mixer, you can use a musical instrument amplifier, such as a guitar amplifier, as long as it has two separate input channels, each with its own volume control. Place two microphones (for left and right channels) by the amplifier's speaker, and plug them into a cassette recorder to record the results on a blank tape.

You will make a one-minute recording. Cue the blank tape past the cassette's leader (the plastic part of the tape at the beginning of the cassette). Set the volume of the white noise source fairly high. Set the volume of the music being played at about an equal volume. Start the recorder, the white noise generator, the music tape, and a stop watch. After ten seconds, slowly begin to turn the volume of the music tape down, but leave the white noise at a constant level. The music fade must be very, very slow. Pace yourself so that at 50 seconds into the song, the volume will be 100 percent reduced. At 60 seconds, stop the tape recorder.

Once you make your test tape, place it in a cassette player with headphones. Have a friend wear the headphones and tell him or her to push the play button. Be sure you have cued the tape up past the leader at the beginning of the tape, so when the play button is pressed, your test recording begins to play. Start timing the instant the tape begins to play. Ask your friend to tell you as soon as he or she hears the music stop playing.

Remember, at 50 seconds into the tape, the music is gone. Does the tape recording end (at 60 seconds) before your friend says the music has stopped? Does your friend say the music never stopped?

Test at least 20 friends, and write down whether each one could correctly identify that the music ended before the tape recording ended.

Results

Write down the results of your experiment.

Conclusion

Come to a conclusion as to whether or not your hypothesis was correct.

Something More

1. Does age have any effect on your results? In other words, do more young people continue to "hear" the music when it is gone than do people over age 50?

2. Does gender have any effect on your results?

3. People may claim to continue to hear what they expect to hear, but what if the music played to them was a song they were *not* familiar with? Would they still claim they were hearing music when it was no longer playing?

Project 6

"Vlip!"

A pet dog responds to sounds rather than understanding the meaning of words

Suggested Entry Categories

- Behavioral & Social
- Zoology

Purpose or Problem

The purpose is to prove that a pet dog who is trained to obey several commands, responds to those commands because of association with the sounds and the action you want from the animal, not because of any understanding of language.

Overview

Pet owners who train their dogs to obey several commands naturally use words in their own language. "Sit," "bark," and "roll over" are words those who speak the English language understand. Although a dog may appear as though it understands the meaning of commands, it is merely the sound of these words that produces the appropriate behavior.

Hypothesis

Hypothesize that a dog can be trained to obey several command words that are not words in any language, proving the animal is merely associating a particular sound with a particular expected behavior.

Materials' List

- Pet dog
- Book on how to train your dog
- Time and patience training the dog to obey several commands

Procedure

Decide on several behavioral responses you want to train your dog to accomplish (sit, stay, run, bark, and so on). Then, make up your own words to substitute for these English words. "Vlip," for example, could be "sit." Make up simple one-syllable words.

If you can train your dog to respond to these made-up words, only the two of you (and no one else in the room!) will understand the commands.

Get a good book on how to properly train your dog to obey voice commands.

Just as most people like to be rewarded for their achievements, so do your pets. Rewarding (giving a hug or a treat) is the best motivation for your pet to learn.

Start by giving three rewards when the dog's response to a command is correct: give a pat on the head, say "Good dog!" and give a food treat. As time goes on, don't give food every time. Eventually, just a pat or hearing the tone of your voice saying "Good dog!" will be sufficient to let the pet know you are proud of it.

Time and patience are needed to train your dog, but it will be fun for both of you. The training will seem more like playing together than work.

Results

Write down the results of your experiment. Document all observations and data collected.

Conclusion

Come to a conclusion as to whether or not your hypothesis was correct.

Something More

1. Children can be bilingual and learn two different words for the same thing. This can happen when one parent or grandparent speaks a different language than the other parent or family members. Can a dog learn more than one command for the same behavior?

2. Pet guinea pigs can be taught to squeal and rattle their cages at the sound of chopping carrots on a wood block, in their anticipation of receiving carrot treats. Can they be trained to get equally excited by a voice command indicating a food treat is coming?

3. Dolphins learn to do tricks by watching their trainers' hand signals. Can dogs learn commands by hand signals only?

Got Salt?

Comparisons of back bay salt content to tide cycles

Suggested Entry Categories

- Environmental Science
- Earth Science
- Chemistry

Purpose or Problem

Comparing salt content in back bay water during high tides and low tides.

Overview

The gravitational pull of the Moon and the Sun creates a daily flow of water toward and away from sea coasts (*high tide* and *low tide*). As water flows toward the coast, the water level along the shore can be seen to rise, and water flows through inlets, filling back bay areas. Hours later, an *ebb tide* occurs, when the water recedes out of the bays and away from the shoreline.

Does this tidal change affect the salt content of the water that accumulates in the back bays? If a significant difference exists between the salt content at high tide and low tide, plants and animals living there would have to be tolerant of these changes.

Hypothesis

Hypothesize that a noticeable difference will occur in the salt content in back bay water depending on the cycle of the tide (high tide, low tide).

Materials' List

- Access to an inlet and bay areas that experience tidal changes, fed by an ocean or a large body of salt water
- Four wide-mouth jars (peanut butter, pickle, or other food containers) of equal size
- Masking tape
- Pen or marker
- About two weeks of waiting time
- Several small twigs
- A sunny window
- Tide chart helpful, but optional
- Possible adult supervision needed

Procedure

The amount of water gathered for each sample and the location the samples are taken from remain constant. The tide cycle is the variable.

For this project, you must have access to an inlet and a back bay that receives tidal

flows from an ocean or a large body of salt water. **When you work around water, make safety your number one concern. Know how to swim, wear a life preserver, and always have a friend or an adult accompany you.**

Gather four clear glass or plastic jars that have wide mouths. Jars of this type include 16- or 18-ounce peanut butter, pickle, or sauce containers. All four jars must be identical.

Place a strip of masking tape on each jar and label each one as to the location and tidal status that identifies the water sample they will contain.

You need to determine the time of high and low tides. Tide tables are often found in local marinas, newspapers, or by listening to a National Oceanic and Atmospheric Administration (NOAA) weather station (weather radios can be purchased at many consumer electronic stores). If you do not have access to a tide table, you can spend a day making note of where the high- and low-tide levels are along bulkheads or other land markings. Throwing a small twig in the water at an inlet and watching the direction it floats tells you whether the tide is flowing in or receding out.

The figure on the top of the next page shows two points where you should collect water samples: one is located in a back bay area, and the other is at the mouth of the inlet, where the bay meets the ocean.

When the tide is just beginning to flow in (just past the time of low tide), fill a jar with water from Point A and one from Point B. Secure lids on the jars to keep the water from spilling as you transport them home.

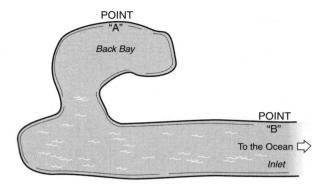

Later, when the tide is just beginning to recede (just past the time of high tide), fill a jar with water from Point A and one from Point B.

[Optional: If you have access to two inlets that feed back bay areas, you can enhance your project by collecting additional samples at points shown in the figure below.]

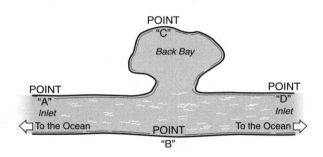

When you get home, place the jars in a warm, sunny window and remove the lids. It takes about two weeks for all the water to evaporate. You can decrease evaporation time

by placing them in an area of increased heat, such as near a heat duct or in an oven at a low temperature. Do not place the jars directly on a stove burner, as the jars are not designed to be exposed to high temperatures.

When all the water has evaporated, screw the lids on again to prevent any further contamination and to keep the contents intact.

Do you see chunks of salt in the jars? Salts are crystals, and one of the characteristics of crystals is their unique shapes. Do the chunks have shapes characteristic of crystals? You may also want to examine the salt chunks under a magnifying glass or microscope.

Results

Write down the results of your experiment. Document all observations and data collected.

Conclusion

Come to a conclusion as to whether or not your hypothesis was correct.

Something More

1. Is there a difference in salt content in water near the surface compared to water at a deeper level? Construct a device that lets you lower a container to the bottom of the bay, and then open the lid to fill it with water. (Use a brick or a heavy object to weigh down the container.)

2. Does heavy rainfall affect the salt content in a back bay?

3. Is there any difference in salt content between the water in a back bay and the water in the ocean that feeds it?

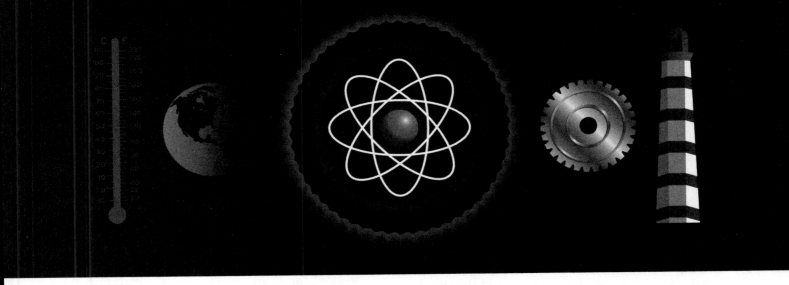

Project 8

In the Ear of the Beholder

The physics and social classification of "noise"

Suggested Entry Categories

- Behavioral & Social
- Physics
- Adaptable to Math & Computers

Purpose or Problem

Defining "noise" can sometimes be explained by the physics of a sound, but it can also be in the mind of the listener, or it can even be determined by the society the listener lives in.

Overview

Some sounds are musical and some are noise. *Music* is usually made of vibrations that are organized and come to our ears at regular intervals. A vibrating string on a guitar or a piano is an example. You can feel the vibrations of your vocal cords by placing your hand on your throat while you sing.

Sounds that make irregular vibrations tend be thought of as noise. Such vibrations are made when a door is slammed shut or a book falls from a desk to the floor.

But, it is not always easy to classify a sound as music or noise. The difference

between music and noise may be in the mind of the listener. Young people's music may be considered noise by their grandparents. Hitting a fence with a stick may be noise, but if you walk along a picket fence and hold a stick against it, the regular repetitive sound may be pleasing, as a drum or other percussion instrument would be.

Society also determines when a sound is musical. Do research into the unusual instruments used in other cultures.

Have you ever listened to someone squeaking as they learn to play the clarinet, or screeching as they learn to play the violin? That's hardly music!

Perhaps we can classify sounds as being more musical if we enjoy them. A sound may even be pleasing at one moment, but not at another. A doorbell may make a harmonious sound, but if it dings at 3 o'clock in the morning and disturbs you while you are sleeping, you won't like it. When music is played softly, it may be enjoyable, but when the volume is turned up to the point where it hurts your ears, the song becomes noise.

The time of day may also affect your feelings about a sound. If you are waiting for a friend to pick you up to go to the movies,

the honk of their car horn is welcomed. But someone honking a car horn in front of your house at 4 A.M. can be disturbing. A doorbell ringing during the day does not have the same alarming affect emotionally as it does if it rings in the middle of the night. When a mother knows why a baby is crying (if it needs a diaper changed or is hungry), her emotions are not the same as when she doesn't know what is wrong.

Some sounds always seem to be pleasant: a babbling brook, the wind rustling leaves through the trees.

We are surrounded by sounds all day long, and it is important that we have quiet times and enjoyable sounds in our daily lives. Too much noise can cause stress and fatigue.

Hypothesis

Hypothesize that in the categorizing of common, everyday sounds as to whether they are pleasant or noise, many different responses will be based on the age group.

Materials' List

- Paper
- Pencil
- Clipboard
- A day of listening
- Ten friends of high-school age
- Ten adults over age 40

Procedure

The list of commonly heard sounds will be constant for all who are surveyed. The age groups of those surveyed will be varied: teenagers and adults over 40.

For one whole day, pay attention to all the sounds you hear. Carry paper, a pencil, and a clipboard to make a list of all the daily sounds around your home, school, and neighborhood. Some sounds you may not have paid much attention to before: for example, toast popping up in a toaster, a door chime, a church bell, popcorn popping, a car horn, the crackling of a fire in a fireplace, the telephone ringing, birds chirping, someone tapping a pencil on a desk, an umpire or referee blowing a whistle during a sporting event, insects buzzing in your ear, the screech of car brakes, the blowing of air across the top of a soda bottle, or someone driving by in a car with your favorite song playing.

Compile a survey sheet with a list of 50 sounds, each followed by a multiple choice selection of Pleasant, Noise, and No Response.

At the top of each sheet, make a place for checking the two age groups: Tees and Over 40. (You can also ask for male or female if you want to do the "Something More" suggestion.)

Use a copy machine to make 20 copies, or use a computer word-processing program or desktop publishing program to create your survey sheet, and print out 20 copies.

Have ten high-school-age friends and ten adults over age 40 complete the survey. Total the results from each group. Compare the responses by each group.

Results

Write down the results of your experiment. Document all observations and data collected.

Conclusion

Come to a conclusion as to whether or not your hypothesis was correct.

Sample Survey Sheet

Age Group: _____ Teens _____ Over 40

Check each sound as to whether that sound is pleasing, "noise", or neither to you.

1. A door bell ringing in the afternoon.
___Pleasant ___"Noise" ___No response

2. A car driving by you as you sit outside your home, with its radio blasting.
___Pleasant ___"Noise" ___No response

3. "Big Band" dance music.
___Pleasant ___"Noise" ___No response

Something More

1. Expand your survey by categorizing your results by male and female, in addition to age. Compare your organized data.

2. Sounds and songs may even bring memories to our minds. Hearing a popular song that was once played heavily on the radio may cause you to remember a special summer or time in your life. When you hear the sound of sleigh bells or a Christmas carol, does a feeling or picture come to your mind about snow falling or the excitement of waiting to open presents with family?

Project 9

Flying in the Wind

Wind velocity at ground level may be different at heights above the ground

Suggested Entry Categories

- Environmental Science
- Earth Science

Purpose or Problem

The purpose is to determine if wind speed is different at ground level compared to 30 or 40 feet above ground.

Overview

The rotation of the Earth and differences in atmospheric temperature give birth to an inexpensive and renewable source of energy ... the wind.

Down through the centuries, wind has been a powerful source of energy that mankind has harnessed to do work. The wind fills the sails of ships and turns the blades of windmills, which once were used to grind grains and saw wood, and today are used for generating electricity.

Studying the behavior of the wind is one of the most important aspects of meteorology, and it leads to a better understanding of weather and weather forecasting.

Is the speed of the wind different at different heights above the ground? Have you ever been sitting on the ground and, while you only felt a slight breeze, you could see the tops of very tall trees swaying in what appeared to be a stronger wind? Are the blades of windmills built up high because it is usually windier up high than it is near the ground?

Hypothesis

Hypothesize that the wind is often stronger at a higher distance from the ground.

Materials' List

- Nine feet of ribbon, 2 inches wide
- One-week period of time
- Several clip-type clothespins
- Use of a high flagpole
- Pencil and sketch pad
- Use of a camera (optional, but useful in making a science fair presentation)
- Possible adult supervision needed

Procedure

The location of the flagpole, the height of each ribbon wind indicator, and the ribbon indicators themselves are constant. The wind speed is the variable.

Get permission to use a tall flagpole that is away from buildings and other structures. Sometimes, local businesses will have high flagpoles for promotion. Your school may also have a tall flagpole.

Cut three 3-foot lengths of ribbon, the kind used for decorative craft work. The ribbon should be about 2 inches wide.

Tie the three pieces of ribbon onto the rope that hoists up the flag. Space the ribbons so that when the rope is pulled up, one ribbon will be at the top, one at the middle, and one at the bottom of the pole.

Every day at the same time for seven days, observe the position of the ribbons. Use a sketch pad to record your observations. The ribbons will give a relative indication of wind speed. The straighter they stand out (parallel to the ground), the stronger the wind speed.

If there is a day when no wind is blowing and none of the ribbons are moving, do not record an observation. Instead, wait until another day when there is enough wind to move at least one of the ribbons.

If stormy conditions exist, do not record your observations. **Being outdoors in bad weather is unsafe, especially during a thunderstorm.**

If the wind is very strong during one of your observation days, and all three ribbons are standing out straight, try adding weight equally to all of them, so they will not all stand out straight. Weight can be added by clipping one or more alligator-type clothespins to each ribbon.

Results

Write down the results of your experiment.

Conclusion

Come to a conclusion as to whether or not your hypothesis was correct.

Something More

1. Compare your ribbon wind indicators at different times of the day: early morning, noon, and dusk.

2. Can you determine any relationship between the strength or direction of the wind and a barometer reading and the type of clouds?

Project 10

Lighter Struts

Making materials lighter, yet still strong enough for the required need

Suggested Entry Categories

- Engineering
- Math & Computers

Purpose or Problem

Determining the safety stress range of a 2×6 piece of balsa wood, while making it lighter in weight.

Overview

Many times a material needs to be very strong, because it will undergo a lot of stress or pressure. Sometimes, a material must be made from a strong substance, like steel, but it may also have a requirement of being as light as possible. Some bones in birds are strong, yet they are hollow to make them light. The struts used in aircraft often have large holes in them to make them lighter, yet they must still be strong enough for the job they are required to do.

Design engineers must know how much force a material can withstand before breaking, and whether or not that material can be made lighter by cutting holes in it, yet still being able to support the weight needed.

While engineers may need a certain material for its strength, the material may be able to withstand much more stress than required. Therefore, they can reduce the mass (weight) of the material by cutting holes in it. A margin of safety must also be included to ensure a safe design. For example, if 2 pounds of stress is to be exerted on a material, you may want that material to be able to withstand 6 pounds before it breaks, giving you a two-thirds margin of safety.

Hypothesis

Hypothesize that you can lighten a piece of balsa wood by cutting holes in it, while still keeping much of its structural strength.

Materials' List

- Plastic gallon jug
- Two-foot-long piece of strong string
- Wooden sawhorse
- Two pieces of balsa wood, 2 inches wide by 1 foot long
- Piece of 2×4 lumber about 12 inches long
- Two pieces of plywood about 1 foot by 6 inches
- Wood screws

- Screwdriver (or electric screwdriver)
- Hand saw (**or a power saw used under adult supervision**)
- Gram weight scale
- Bathroom scale or scale to measure pounds
- Utility knife
- Pitcher of water
- Ruler
- Possible adult supervision needed

Procedure

The wooden box device that holds the balsa wood in place is held constant, as is the water jug device for adding weight to stress the balsa wood. The mass of the piece of balsa wood is the variable.

Cut two pieces of thin balsa wood, 2 inches wide by 6 inches long, and set these strut-like pieces aside.

Cut two pieces of 2×4 lumber in 6-inch lengths. Cut two rectangular pieces of plywood into pieces 1 foot wide by 6 inches deep.

Using screws, attach one piece of plywood to the top of a wooden sawhorse. Using that as the bottom piece, make a rectangular box by using the 2×4 pieces for

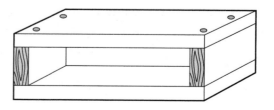

sides and the remaining piece of plywood for the top. Leave two sides of the box open.

Using a utility knife, cut a small V-shaped notch on one of the long sides of the balsa wood near the end of the wood.

Stand the balsa wood vertically and insert one end into the opening of the box to a depth of 1 inch. Position the balsa wood so the end with the notch in it is outside the box and facing upward.

Tie both ends of a 2-foot-long piece of strong string to the handle of an empty plastic gallon jug, making a loop. Hang the jug from the balsa wood by placing the loop of string in the notch.

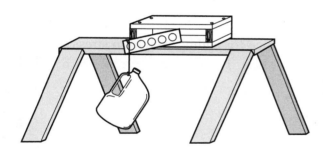

Slowly add water to the jug until the balsa wood breaks. Weigh the water in the jug and record this weight. (If the weight of a full jug is not enough to break the wood, tie a second jug to the first one and begin adding water to it.)

Take the second piece of balsa wood and cut a series of holes in the wood, each spaced at equal distances. Each hole should be 1 inch in diameter.

Perform the same weight test. Record the amount of weight needed to break the "swiss cheese" balsa wood strut.

Engineers need to know weights, percentage relationships, strengths, and other important factors about a material. Use the data you have determined by experimentation to compile stress data on the 2×6 pieces of balsa wood. The fact sheet you compile should include these figures:

- Weight of the solid balsa wood piece (use a gram weight scale):_____

- Weight of the lightened "swiss cheese" balsa wood piece:_____

- Breaking weight of the solid balsa wood piece: _____

- Breaking weight of the lightened "swiss cheese" balsa wood piece: _____

- Percent lighter the lightened strut is to the solid strut (the weight of the lightened piece divided by the weight of the solid piece, times 100 for percent):_____

- The percentage of weight the lightened piece breaks at compared to the solid piece (breaking weight of the lightened piece divided by the breaking weight of the solid piece, times 100 for percent):_____

- The maximum weight the lightened piece can safely support, the safety margin being two-thirds beyond what is required (breaking weight of the lightened piece divided by one-third):_____

Results

Write down the results of your experiment. Document all observations and data collected.

Conclusion

Come to a conclusion as to whether or not your hypothesis was correct.

Something More

If the balsa wood was twice as thick, would it be able to hold twice the weight? Laminate two pieces of balsa wood together with wood glue. Let dry and test.

Stock Up

Concepts of stock market investing

Suggested Entry Category

- Math & Computers

Purpose or Problem

The purpose is to serve as an introduction to stock market investing, and to develop an enthusiasm for saving and investing.

Overview

Every time you buy something, whether it is a can of soda, a pack of gum, or a pair of sneakers, you are helping a company grow and increase its earnings. Many big companies are *publicly traded*, that is, individual investors can buy stock in that company and actually own a small piece of it. When a company does well, the stockholders may benefit from the company's prosperity by receiving a *dividend* (a cash payment) or by the value of their stock increasing.

There is a lot to learn about investing in the stock market, but it is something all of us should do during our lifetime. A portion of our hard-earned money should be at work in the stock market, earning us even more money. We must learn ways to get the money we earn to grow. Putting money in a passbook savings account or a bank certificate of deposit (CD) is a common way to invest. Although these forms of investments are very safe (there is little chance you will lose your principal), they generally do not give a very high return on your investment, perhaps paying only 2 percent to 6 percent annually. The stock market has always been a place where an individual can get a much higher return on the money invested. The catch is this: although the stock market has always performed well over time, it can suffer temporary drops, and investors can lose the money they have invested. For that reason, many people invest in a mutual fund, where a professional money manager studies companies and their stocks, and makes buying and selling decisions for you and many other people who have their money in the mutual fund.

However, there is a thrill of picking a stock yourself and watching it on a daily basis. While you should have a portion of your money invested long term (10 to 20 years) in the stock market, it can be exciting and often profitable by "playing" the market for the short term. To do this will require a little time every day studying the financial newspapers and watching the financial news on television, and using the Internet to find company news and track your stocks.

Two developments took place in the last three years of the 1990s that enabled the average person to easily and cheaply get into the stock market. The first was the explosive growth of the Internet and computers, enabling almost every home to be able to afford a computer and be connected to the Internet. The second development was the appearance of high-discount Internet brokers. Previously, an investor might have to pay $100 or more in commission fees to buy stock. Using the Internet, a transaction can cost as little as $8, making it very affordable to the masses.

In this project, you gain valuable experience learning about the stock market. You will do paper trades (pretend trades as opposed to trading with real money). We hope this will give you a better understanding and insight into the stock market, and make you excited about saving and investing!

Hypothesis

Hypothesize that given an imaginary $10,000 to invest, you can select stocks to buy and sell, and build your initial investment by 10 percent within three months.

Materials' List

- Computer with an Internet connection
- Daily financial newspaper (*The Wall Street Journal* or *Investor's Business Daily*)
- One or two books on an introduction to the stock market

- Three months' time
- Calculator
- Paper and pencil

Procedure

Go to a bookstore or your local library and read one or two books on an introduction to the stock market. Become familiar with some of the terms you encounter.

Pick up a copy of a daily financial newspaper (*The Wall Street Journal* or *Investor's Business Daily*) and browse through it.

In the project, the initial capital invested is constant. The stocks in your portfolio (you can buy and sell them anytime during the three-month period) and the fluctuations of the stock market are variables.

These are the rules of our short-term trading portfolio:

1. You have $10,000 initially to invest. If any of your stocks increase in value and you sell them at a profit, you can use the extra money to buy shares of another stock.

2. You can buy and sell any stock at any time. However, assume there is a commission fee of $10 for every transaction. When you buy a stock, add $10 to the total cost. When you sell a stock, subtract $10 from the profit you receive.

3. Any money left over from the $10,000 that is not invested in stock is assumed to make 4 percent annual interest, as it is swept into a money market account by your broker. Calculate the daily income from that uninvested money and add it to your three-month total. (4 percent divided by 365 days in a year is about .01 percent per day earned on the uninvested balance.)

4. At the end of three months, sell all your stocks. Total their value and add any other profit you made from the selling of stock during the three-month period. Add money market interest. Subtract any losses you incurred by selling a stock that was underperforming.

5. You must maintain at least six stocks in your portfolio at all times. This will give you *diversity*, which lowers your risk of losing money by not "putting all your eggs in one basket," in case one stock takes a big drop.

To select the initial six or more stocks to begin your portfolio, you need to make a list of potential stocks to invest in. Start by writing down the names of companies whose products you like or use. Do you like to collect Disney toys? Do you like to drink Pepsi Cola? Is McDonald's your favorite hamburger stop? Is your hobby surfing or rollerblading?

Once you have a list of about ten companies, do research on each one. The Internet has hundreds of free web sites where you can get information on a company, including a profile, fundamentals (the highest and lowest stock price for the year, Price to Earnings (P/E) Ratio, number of average daily shares traded, and so forth), and a price chart showing the history of the stock.

Studying a chart is called *technical analysis* and, although a stock's history is no guarantee of what will happen in the future, it often gives a good indication of which direction the stock is likely to be headed. If a chart shows a stock price has been dropping for the last three months, it is probably a company you want to avoid.

The Internet, daily newspapers, and financial TV programs are good sources for hearing news about your companies. This requires daily monitoring. If a cold winter is expected, that might be an opportunity to invest in coat manufacturers. If a company is being sued, whether it is in the right or not, this can put a drag on the stock until the matter is settled.

The whole market goes up and down, and it can carry stocks with it. The stock market is controlled by perceptions of the investment "crowd." The least little thing can drive the market up or down. If the President stubs his big toe, the market may drop. But in such cases, it will probably bounce back up. During times when the whole market takes a dip, don't panic and sell all your stocks. As long as there is no change to the fundamentals of your individual companies, stay the course.

At times, there will be important news on your companies that will affect their price. One of the biggest factors that affects the price of stock is quarterly earnings, and you should pay close attention to earnings' estimates.

As you follow your stocks daily, you can learn about the factors that affect a stock's price: earnings, selling off a division of the company, acquiring other companies, announcing a stock split, and expanding overseas operations.

You can also get ideas for your initial portfolio list by reading financial newspapers and magazines. Also, be observant of products around you. What brand of shoes are most of your friends buying? Where are they buying their school clothes? Who makes your favorite computer games?

Results

Write down the results of your experiment.

Conclusion

Come to a conclusion as to whether or not your hypothesis was correct.

Something More

1. Generally, financial advisors recommend buying a stock and holding it for a long time to get the most benefit. Continue your project for six months or even one year. Do your stocks do better over time?

2. Get a book on an introduction to the options market. Set up a portfolio of stocks and paper trade writing covered calls, buying Call options, which represent the investor's right to buy stock, and selling Put options, which represent the investor's right to sell stock.

Project 12

A Better Burger

Comparing the fat content in different grades of ground beef

Suggested Entry Category

• Chemistry

Purpose or Problem

The purpose is to compare the fat content of different grades of ground beef. Too much fat in our diet may be unhealthy, especially if we do not get enough exercise. We should be aware of fat in the foods we consume, so we can make healthier choices when purchasing such items as ground beef (lean or extra lean) and milk (whole or skim).

Overview

Fats are substances found in animals and some vegetables. Fats are used by the body for energy. When the body's demand for heat increases, as it does during the winter or in cold climates, more fat is required by the body. Eskimos, for example, consume a great amount of fat in their diet. When more fat is eaten than the body currently needs for growth or energy, the fat is stored in tissues.

Fats are made up of carbon, hydrogen, and other elements. The carbon and hydrogen give fats their capability to give off huge amounts of heat. Have you ever cooked chicken still in its skin on a barbecue grill? If so, you probably noticed that when the chicken fat fell into the fire, it caused a flare-up of flames.

Hypothesis

Hypothesize there is significantly more fat in standard ground beef than in ground beef labeled extra lean.

Materials' List

- Adult supervision (**exercise caution around a hot stove**)
- Package of ground beef
- Package of lean ground beef
- Package of extra lean ground beef
- Spoon
- Small cooking pot
- Kitchen gram-weight scale
- Three paper plates
- Water
- Measuring cup
- Clock or watch
- Pencil and paper
- Use of a kitchen stove
- Use of a refrigerator

Procedure

Fats can be liquefied by heat. When you fry bacon in a pan, you have no doubt noticed all the fat that appears in the pan. Heating bacon in a microwave oven yields the same result.

The mass of each meat patty, the cooking time, and the amount of water used for boiling are all constant. The fat grade of ground beef is the variable.

Obtain three small packages of ground beef, one simply labeled "ground beef," one labeled "lean," and one labeled "extra lean."

From each package of meat, mold a hamburger-shaped patty. Use a small kitchen scale to make sure each patty weighs the same.

Place the first patty in a pot and put it on a stove burner. Add one or two cups of water as needed to completely cover the patty. Record the amount of water added.

Note the time on a clock or watch. Turn the burner on its highest heat setting. Bring the water to a boil. Carefully break up the patty into small pieces as it is boiling, so the boiling water can reach all parts of the meat. **Use extreme caution when working around boiling water.** The heat will extract the fat from the meat, and the fat will rise to the top of the water.

After several minutes of boiling, turn the burner off and note how much time has passed on the clock.

Weigh a paper plate on a kitchen gram weight scale and record the weight.

Place the pot and its contents in a refrigerator. As the fat cools, it will *coagulate* (change into a thickening mass). After the coagulated fat has cooled for several hours, scoop it off the top with a spoon and place it on the paper plate. When all the fat is on the plate, weigh it. Subtract the weight of the paper plate (the *tare* weight) to determine the weight of the collected fat.

Clean out the pot and repeat the process for each of the remaining two grades of ground beef. Use an equal amount of water to boil each patty and boil each for the same length of time.

Compare the weights of the fat collected from each grade of meat. Is the amount of fat significantly less in the extra lean grade than the other grades?

You know the original weight of the patties and the weight of the collected fat. What percentage of each patty was fat? Divide the fat weight by the patty weight and multiply by 100.

Results

Write down the results of your experiment. Document all observations and data collected.

Conclusion

Come to a conclusion as to whether or not your hypothesis was correct.

Something More

1. Compare the fat content in meat mixtures (beef and pork are often sold together).

2. Ground beef that contains more fat may not be as healthy, but is it tastier?

Project 13

Caught in the Spotlight

Devising an insect-collection device, and then evaluating the nocturnal insect population in your area for health hazards

Suggested Entry Category

- Medicine & Health

Purpose or Problem

Insects can carry and transmit diseases that are harmful and even life-threatening for humans. We should be aware of the types of insects in the area we live in and know how to protect ourselves from them.

Overview

To many of us, insects and bugs are merely pests, getting in our food at a picnic or giving a bite that makes us hurt or itch. However, some insects can be very harmful to humans, even causing death. Many people are allergic to bee stings. Ticks can carry Lyme disease and Rocky Mountain spotted fever. Mosquitoes can spread deadly diseases, including malaria and yellow fever. It is important to know what species of insects live in your community.

This project attempts to collect a representative sample of the type of insects in your area. First, several devices for collecting insects will be constructed and tested. Then, the successful collecting device will be used to gather insects, which can then be studied, identified, and researched to see if they might be a potential health hazard to humans.

Respect for all forms of life is imperative. Therefore, only a short period of time will be allowed to collect insects to have a minimal effect on the environment and to prevent the needless loss of insect life.

Hypothesis

Hypothesize which of the four insect-collecting devices you have constructed will collect the most insects during a given amount of time. Also, hypothesize that you will collect more than ten different organisms.

Materials' List

- Book on *entomology* (a branch of zoology that deals with insects) to help you identify the insects you collect
- Four 2-liter plastic soda bottles
- Four wide-mouth jars
- Two flashlights
- Clock or watch
- Outdoor area away from bright lights in the evening
- Black construction paper
- Adhesive tape
- Magnifying glass

- Scissors
- Marker pen
- Possible adult supervision needed

Procedure

First, we must construct a device that will collect a sampling of nocturnal insects. To do this, we will construct four different devices and test them to see which is the most successful. Cut the top and bottom off four 2-liter plastic soda bottles, leaving just the hollow cylinders. **Use caution when working with sharp scissors.**

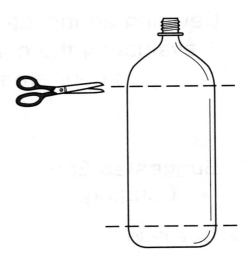

Find an outdoor location around your home or neighborhood that is safe, but away from strong lights. On the ground, set up the four plastic cylinders. Place flashlights inside two of them, with the beam facing straight up.

On top of each cylinder, set a wide-mouth jar. Keep the lids screwed on the jars.

Fill one of the jars that is over a flashlight half full of water. Next, fill one of the jars that does not have a flashlight underneath half full of water.

Using black construction paper and adhesive tape, wrap each cylinder and jar with the paper, so light from the jars that have a flashlight underneath them will only shine out of the top.

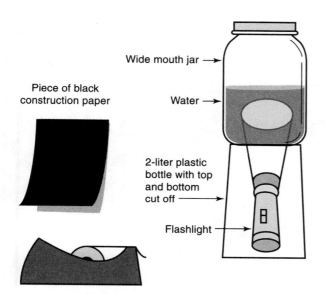

Piece of black construction paper

Wide mouth jar →

Water →

2-liter plastic bottle with top and bottom cut off →

Flashlight →

When it's dark, turn on the flashlights and remove the lids from the jars. The size of the jar openings, the location, the time of night, and the period of time the collectors are exposed are the constants in the experiment. The variables are light and water.

After one hour, screw the lids back on the jars, and then turn off the flashlights. With a marker, label the lids of each jar as "Dry, no flashlight," "Dry, with flashlight," "Wet, no flashlight," and "Wet, with flashlight." Then, take the jars inside.

Which device collected the greatest quantity of insects? Was your hypothesis correct?

Using a book on insects, identify the insects collected. Research more about each insect and create a fact sheet on each, including information on whether or not they are harmful to pets or humans, and, if so, why they are harmful.

Results

Write down the results of your experiment. Document all observations and data collected.

Conclusion

Come to a conclusion as to whether or not your hypothesis was correct.

Something More

1. Did the unlit dry collector attract different insects than the lit wet collector?

2. Construct several of the insect-collecting devices that were the most successful at attracting insects. Give one to each of several friends who live in different locations in your community. Have them all expose their collectors for the same period of time during the same evening. Examine both the quantity and type of insects collected. Does one location have a higher insect population than another? Why? Are the same types of insects found at all locations?

3. Are different types of insects active at different times of the night? Collect insects just after dusk and just before dawn (when it is still dark).

Sweet Treat

The behavior of ants toward natural and artificial sugars

Suggested Entry Categories

- Behavioral & Social
- Chemistry
- Medicine & Health
- Environmental Science
- Zoology

Purpose or Problem

If an organism is attracted to a food source that is not nutritional and changes its diet from eating the type of food it needs to a food that is not nutritional, its health may be put at risk. For example, people sometimes throw stale pieces of white bread out in their yard for birds to eat. But, white bread does not have the nutrition birds need, and they may fill their bellies with a less-than-"good" food.

Overview

Almost everyone enjoys eating sweet foods. Although sugar is the most commonly used sweetener, it has been associated with health problems, including tooth decay, obesity, and hyperactivity in children. People with health problems related to sugar could be diabetic or they might be trying to lose weight, and they're looking for sugar substitutes to use in their foods.

Today, two popular artificial sweeteners are sold in grocery stores, namely saccharin and aspartame. Neither of these sweeteners has any nutritional value, whereas natural sugar is a carbohydrate that provides energy for your body. Artificial sweeteners are not used by the body and pass through unchanged.

The natural sugar found on your kitchen table is called *sucrose*, and its chemical symbol is $C_{12}H_{12}O_{11}$.

Aspartame is 200 times sweeter than sugar, and saccharin is 500 times sweeter. Will ants be attracted to these useless food sources because they are sweeter? Or, will nature prevail and their instinct be able to detect that the false sugars are nutritionally empty?

Hypothesis

Hypothesize that the instinct in ants is intelligent enough that when offered natural sugar (which has nutritional value) and artificial sweeteners (which have no nutritional value), the ants will take the natural sugar, even though the artificial sweeteners taste hundreds of times sweeter.

Materials' List

- Table sugar
- Brown sugar
- Saccharin artificial sweetener
- Aspartame artificial sweetener
- One-fourth teaspoon measuring spoon
- Magnetic compass
- Anthill with active ants
- Wooden ice-pop sticks or tongue depressors
- Ruler
- Pen or felt tip marker
- Three days of time
- Stiff piece of cardboard
- Scissors

Procedure

Locate an active anthill in an area where you can safely make observations from a distance of several feet. The terrain around the anthill should be fairly consistent. Do not use an anthill that is bordered on one side by grass and on the other side by a sidewalk. Be sure grass or soil surrounds the hill out to several feet.

Observe the anthill for one day to learn when the ants are most active: early morning, late morning, afternoon, early evening, and so forth.

Once you discover an active time for the ants, set up four piles of sugar and sugar substitutes as explained in the following, just prior to that time on the next day.

Using a magnetic compass, locate North, East, West, and South directions, with the anthill at the center. With a pen or felt tip marker, write NORTH, EAST, WEST, and SOUTH on each of four wooden ice-pop sticks or tongue depressors. With a ruler and zero at the center of the anthill (be careful not to touch or disturb the anthill), place the stick marked NORTH six inches from the center and to the north of the anthill. Push the stick into the ground, so it stands vertically as a marker.

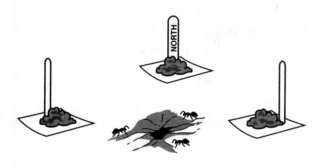

Similarly, measure and push into the ground direction-identifying sticks six inches to the east, south, and west of the mound.

Cut four small squares (about 2 or 3 inches square) out of a piece of stiff cardboard or oak tag. Lay one square in front of each of the ice-pop sticks.

On the piece of cardboard in front of the North stick, use a level ¼ teaspoon of sugar to make a small pile of table sugar.

In front of the East stick, make a pile of brown sugar on the cardboard, again using a level ¼ teaspoon of brown sugar.

Similarly, place saccharin and aspartame piles at the South and West markers.

The piles are placed on pieces of cardboard, so at the end of the day, they can be removed.

Observe the ants as closely as possible, but do not get so close as to affect their behavior.

Are the ants attracted to any of the piles? If so, which ones?

On the third day, place the pieces of cardboard with the piles by the marker sticks, but this time rotate them, so the sugar is east of the anthill, the brown sugar is located to the south, and so on. Observe the ants' behavior. Are they still attracted to the same piles, even though the piles are in a different place?

If they are, then we can be assured that the location of the piles was not a factor in determining which pile the ants were attracted to, thus eliminating the variables of terrain (uphill, downhill, easier path to navigate, and so forth), the position of the Sun, and the location from the hill (north, east, and so forth).

Results

Write down the results of your experiment. Document all observations and data collected.

Conclusion

Come to a conclusion as to whether or not your hypothesis was correct.

Something More

1. Move the piles of sugars five or six feet from the anthill opening. Does this project then yield a different result?

2. How do ants behave when offered other natural sweet substances? Honey and maple syrup, for example, are often used as a substitute for sugar in baking and other food preparations.

3. If the ants eat the nonnutritious sweeteners, how does it affect other organisms that eat the ants?

4. Large marking pens have a very strong smell. Will the marker odor cause a change in behavior of the ants?

C, a Fantastic Vitamin

The effect of boiling on the vitamin C content of carrots

Suggested Entry Categories

- Chemistry
- Medicine & Health

Purpose or Problem

We must have vitamin C to live, but our bodies do not produce it. Therefore, we must get it from what we eat. The problem is this: even if we eat a food that we know contains vitamin C, our preparation to eat it (boiling the food, for example) may cause vitamin C to leave a food.

Overview

Vitamin C (its chemical name is ascorbic acid) is one of the most important vitamins our bodies need, not only to stay alive, but also to keep in optimum health. Vitamin C is necessary in the formation of *collagen*, which is used to maintain skin, bones, and supportive tissue. Vitamin C strengthens the immune system and aids in healing wounds.

The process of cooking may cause vitamin C to be lost in foods, so that when we eat them, we are not getting this much-needed nutrient.

Is it healthier to eat a carrot raw or boiled regarding vitamin C? In this project, we will boil carrots in water and test the water for vitamin C content before and after the carrots have been boiled in it. If no vitamin C is present in the water before boiling, but is present afterwards, then the cooking process has removed vitamin C from the carrots, thereby making them less nutritious to eat.

Hypothesis

Hypothesize that when carrots are boiled in water, vitamin C will be lost in the carrots. This will be evidenced by an increase in the vitamin C content of the water in which the carrots were boiled.

Materials' List

- One carrot
- Two large test tubes
- Teaspoon measure
- Distilled water
- Corn starch
- Vitamin C tablet (250 milligram)
- Measuring cup
- Funnel
- Cooking pot
- Use of a stove burner
- Iodine
- Eyedropper
- Spoon
- Vegetable peeler
- Possible adult supervision needed

Procedure

When working around a hot stove, use caution. Also, do not put iodine in your mouth or bring it in contact with anything edible, because iodine is poisonous.

Pour one cup of distilled water in a cooking pot. Add ½ teaspoon of cornstarch. Heat on a stove burner. Stir until dissolved, and then set aside to cool.

Pour one teaspoon of this solution into one cup of distilled water. Add four drops of iodine and stir. The solution will turn dark blue. This is our vitamin C test solution. When vitamin C is added to this solution, the dark blue coloring will vanish.

Prove this solution will work in detecting vitamin C by dropping a 250 milligram tablet of vitamin C into the test solution. As the tablet dissolves, the water will instantly become clear.

Now that you have proven the test solution will detect the presence of vitamin C, make another batch of the test solution by pouring one teaspoon of the cornstarch solution into one cup of distilled water and adding four drops of iodine.

Place two large test tubes in holders side-by-side. Pour distilled water into one test tube until it is ¾ full. A funnel is helpful. Using an eye dropper, add drops of the vitamin C test solution until you begin to see the water turning slightly blue. It may take 50

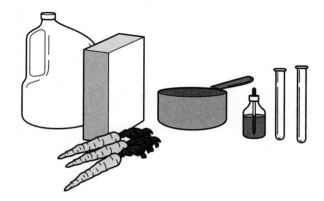

or 60 drops before a color change can be detected. Write down the number of drops added when you first see the blue color.

Cut a carrot into small pieces or use a peeler to slice it into long, thin slivers. Place the pieces in a cooking pot with one cup of distilled water. Bring to a boil for several minutes, and then remove the pot from the heat and let it cool.

Position the funnel in the mouth of the second test tube and pour in the water from the carrot pot until the test tube is ¾ full.

With an eyedropper, add drops of the vitamin C test solution to the carrot water. If vitamin C is present in the water, the water will not show any signs of turning blue when the same number of drops are added that were added to the water before carrots were cooked in it. Can you fill the rest of the test tube and still not see any shades of blue? If so, a significant amount of vitamin C is present. The carrot lost some of its nutritional value to the water.

Results

Write down the results of your experiment. Document all observations and data collected.

Conclusion

Come to a conclusion as to whether or not your hypothesis was correct.

Something More

1. Can any more vitamin C be extracted from the carrot pieces? Place the carrot pieces in a fresh cup of distilled water, reheat, and then retest for the presence of vitamin C in the water. If more vitamin C came out, then some vitamin C was still in the carrot, even though some was lost initially to cooking.

2. Does steaming vegetables retain more vitamin C in the vegetable than boiling them? If so, people who like their vegetables cooked, but who are concerned about maintaining a high level of vitamin C, could steam their vegetables rather than boil them.

3. Can you make a tasty drink from the boiled carrot water? If so, it would be nutritious, because the boiled water contains vitamin C from the carrots.

Zenith Is Not a Radio

Comparing the Sun's daily zenith to the time between sunrise and sunset

Suggested Entry Categories

- Earth & Space
- Math & Computers

Purpose or Problem

The purpose is to determine if the Sun's highest altitude in the sky during the day is at the time exactly halfway between sunrise and sunset.

Overview

The Sun illuminates half of the Earth all the time. But the length of daylight is not half of a day (12 hours) every day. In fact, the length of daylight changes daily and varies at different locations on the Earth.

From about December 22nd to June 21st, the length of daylight increases daily in the Northern Hemisphere, and it decreases from June 21st to December 22nd. This is because the Earth does not spin on its axis in the same plane as it orbits the Sun. The Earth is tilted toward the orbiting plane at a 23½ degree angle.

This tilt not only accounts for why the length of daylight varies, it also gives us a changing of seasons from spring, summer, fall, and winter.

The Sun rises in the east and sets in the west, but the path it travels across the sky changes as the seasons change. The *altitude* (its height above the horizon) of the Sun reaches a higher angle during the day in the summer than in the winter.

Does the Sun reach its highest point in the sky (its *zenith*) during the day at the time that is halfway between sunrise and sunset for that day?

Knowing the position of the Sun at all times during the day throughout the year at a particular location is important to architects who design buildings. They need to know how much sunlight will enter through windows because this will affect their designs for lighting, heating, and air conditioning.

Hypothesis

Hypothesize that the Sun reaches its zenith during the day at the time that is equally between sunrise and sunset.

Materials' List

- 1×1 square piece of plywood
- Large protractor
- Pencil
- Carpenter's level
- Modeling clay

- Sunny day
- Time of sunrise and sunset for that day (available from a daily newspaper, the *Farmers' Almanac*, your local radio station, or a National Oceanic and Atmospheric Administration (NOAA) weather radio)
- Possible adult supervision needed

Procedure

First, be sure you never look directly at the sun! Next, find the time of sunrise and sunset for the day you do this project. These times can often be found in a local daily newspaper, from an almanac, from the news on a local radio station, or from a NOAA weather radio.

Then, calculate the total number of minutes between sunrise and sunset. Divide that number by two. Convert the answer to hours and minutes. Add that to the time of sunrise to arrive at a time that is midpoint between sunrise and sunset.

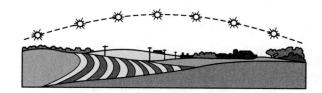

Now, mount a protractor on a small piece of plywood, so it stands perpendicular to the board. Modeling clay can be used to secure the protractor.

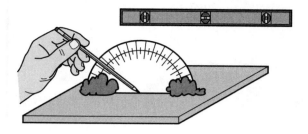

Next, set the protractor device outside in an area that receives unobstructed sunlight all day. Lay a carpenter's level lengthwise, and then widthwise, to level the board. Pile a little sand or small stones under the board to level it.

The reference point of zero on the protractor will be held constant (the plywood and protractor will be kept level horizontally with the ground). The movement of the Earth is the variable in this project.

About two hours before midday, begin measuring the angle of the Sun's altitude by placing the point of a pencil by the base at the middle of the protractor, and raising the pencil up or down until no shadow of the pencil is cast. ***Remember, do not look directly at the sun during this project.*** Check the angle of the pencil by reading the increments on the protractor.

Make these measurements at ten-minute intervals. Write down the angle of the Sun's height.

Continue making measurements until one hour after the Sun's angle begins to decrease.

Did the Sun's peak height occur at about the time you calculated the midpoint between sunrise and sunset?

Results

Write down the results of your experiment. Document all observations and data collected.

Conclusion

Come to a conclusion as to whether or not your hypothesis was correct.

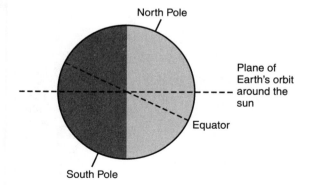

North Pole

Plane of Earth's orbit around the sun

Equator

South Pole

Something More

1. Is the Sun always at zenith at 12 o'clock noon?

2. Determine the *azimuth* (the angle between sunrise in the east and sunset in the west) of the Sun. Thumbtack a piece of white paper onto a piece of plywood, and slightly hammer a nail in the center through the paper and into the plywood, just deep enough for the nail to hold. At sunrise, draw a line tracing the shadow cast by the nail. Just before sunset, when the shadow cast by the nail is about to disappear, draw a line tracing the shadow. Measure the angle. Repeat this once a week for a month. Does the angle change? Is the angle increasing or decreasing in size? Which season are you headed toward? Does the peak height of the Sun in its path across the sky change from week to week?

Bold Mold

Environment affects the rate at which food spoils

Suggested Entry Categories

- Biochemistry
- Environmental
- Microbiology

Purpose or Problem

The purpose is to identify several environmental factors in the home that affect the rate at which mold grows on food, spoiling it. By controlling how food is stored, we can extend its edible life.

Overview

Have you ever been hungry for a sandwich, only to discover that the last two slices of bread in the house have green mold on them? Yuck! Finding mold on your bread in the bread box, mold on your cheese in the refrigerator, or mold on your peaches in the fruit basket can be unpleasant and irritating.

Mold, and another member of the fungi kingdom, mildew, can even attack books. You may have picked up a book that has been stored in a damp basement and found it covered with a powder-like substance.

Molds grow from spores that can travel through the air. Unlike green plants, which can make their own food by using chlorophyll and energy from sunlight, mold must get its nutrients from the food it grows on. Mold breaks down the food it is growing on, causing the food to rot. While finding mold on your sandwich may be unpleasant, rotting is a natural and necessary process of nature.

To keep the foods we buy edible for as long as possible, there are certain things we can do at home to preserve freshness. Will keeping foods from being overly moist or at a cool temperature prolong their viability?

In many instances, food producers add chemicals to their products to preserve freshness. Food additives called *mold inhibitors* destroy microorganisms. The U.S. government allows low levels of certain food additives to be placed in our foods. These chemicals include monosodium glutamate, ethyl formate, sodium benzoate, sodium and calcium propionate, sodium nitrite and nitrate, sorbic acid, and sulfur dioxide.

Natural preservatives include salt and sugar, which bind with the water in foods, preventing the water from being available for microorganisms to feed on. Another method is subjecting food to smoke. You may have read that in early times, meats and fish were smoked to preserve them. Today, foods are usually smoked to give them a distinct flavor, not because it helps preserve them.

Determine what factors you can control at home that help prevent mold growth on common foods. Light, temperature, and humidity are factors we will test.

Hypothesis

Form a hypothesis as to which factors (light, temperature, humidity) or combination of factors play a part in affecting the rate of mold growth.

Materials' List

- Six leaves of lettuce
- Six slices of homemade or bakery fresh bread (but not bread from the supermarket)
- Six small pieces of meat
- Six small pieces of a block of cheese (for example, sharp cheese or Monterey Jack)
- Six peaches
- Spray bottle filled with water
- Box of sealable plastic food storage or sandwich bags
- Use of a refrigerator
- Masking tape
- Felt-tipped marker

Procedure

Gather the foods in the materials' list. Each food will be placed in a different environment. Six pieces of each type of food are needed to test the environmental conditions.

Place each of the foods in a separate sealable plastic food-storage bag: lettuce, bread, cheese, meat, and peaches. Before sealing the bags, spray a light water mist into

one of the two bags for each food, to give extra moisture. Stick a piece of masking tape on each bag that moisture has been added to, and write "moisture added" on the tape with a felt-tipped marker.

Put two bags of each food in a refrigerator. Place two bags of each in a well-lit, out-of-the-way location where the temperature constantly stays between 68 and 72 degrees Fahrenheit. Place two bags of each in a completely dark location (a dresser drawer, closet, or another dark location) that stays at room temperature.

The environments, then, are

- Exposed to light with only the moisture present in the food

- Exposed to light with additional moisture present

- Kept in the dark with only the moisture present in the food

- Kept in the dark with additional moisture present

- Kept in the dark at a lower temperature with only the moisture present in the food

- Kept in the dark at a lower temperature with additional moisture present

Once a day, inspect each food. Write down your observations.

After several weeks, thoroughly inspect all the foods one final time. Do not open any of the sealed bags. When the project is completed, discard the bags in the trash without ever opening them. Breathing mold spores may irritate some people.

Did any of the foods develop mold? If so, which one(s) did the most mold grow on? Which environment was the best at preventing mold? Which encouraged mold growth the most?

If no mold grew on any of the foods, it could be because no spores were present on the foods, or that the foods had been chemically treated with mold inhibitors. Generally, anything homemade would not contain food additives. Try baking a cake from scratch and using it as the test food.

If you were unsuccessful at finding mold on any of the foods, try setting foods out in the open for several days before sealing them in the bags and placing them in their test environments. You may even want to wait until mold begins to form before placing them in the test environments, and then monitor the growth rate of the molds to see the effect of each environment. If left out in the open, spray the samples you are giving extra moisture once or twice a day because the moisture will evaporate.

Results

Write down the results of your experiment. Document all observations and data collected.

Conclusion

Come to a conclusion as to whether or not your hypothesis was correct.

Something More

1. What color is the mold on different foods (bread, cheese, peaches, and so forth)? Do different molds have different textures?

2. Will direct sunlight affect the growth of mold? If so, what is it about sunlight that stunts (or promotes) the growth (visible light, ultraviolet light, infrared, and so forth)?

M&M's Ring Around the World

Determining the validity of sample size

Suggested Entry Category

- Math & Computers

Overview

A lot of investigating can be done with several bags of M&M's candies! And when you finish your investigations, you can eat them! (But, do not eat them if you are allergic to chocolate or can't tolerate sugar!)

Just think of all the questions you can pose:

- How many blue candies are in a bag of M&M's?

- What percentage of each color is in a bag?

- Are the same number of each colored candy found in each bag, or are their quantities random?

- Can you predict how many of each color will be in a bag before it is opened? (You can if the colors are not random.)

- Some people believe there are always more brown candies than any other color. Is this true?

- Does each bag contain exactly the same number of candies?

- Does each candy weigh the same, or are some bigger than others?

Maybe you should eat a few candies before exploring the answers to these questions, just to be sure you have plenty of energy!

Hypothesis

Hypothesize that four bags constitute a large-enough sample size to know whether or not you can predict how many of each color candy will be in a bag before it is opened.

Materials' List

- Four bags of M&M's candies
- Gram weight scale
- Coat hanger
- Three paper clips
- Piece of string
- Use of a door jam

Procedure

Purchase four bags of M&M's candies (or similar bags of multicolored candies). With a pencil and paper, draw a table with column headings for BAG #1, BAG #2, and BAG #3. The number of each colored candy will be recorded under each bag. If you know how to use a spreadsheet program on a computer, set up the table on the spreadsheet.

COLOR	BAG #1		BAG #2		BAG #3	
	Quantity	%	Quantity	%	Quantity	%
Blue						
Brown						
Yellow						
Orange						
Green						
Red						
TOTAL						

Open a bag of M&M's candies. Count the number of each colored candy, and write the number down on the table under the column labeled BAG #1.

In the same way, open a second bag, count and write down the number of each color under the BAG #2 heading. Finally, do the same with the third bag. Add each column and write down the total number of candies in each bag.

Looking at your table of data, you can now answer all the questions we posed.

- Percent—*Percentage* is a measure of a part of something to the whole thing. It is expressed in hundredths. *Percent* comes from the Latin words *per* meaning by, and *centum* meaning one hundred. The symbol for percent is %. To find the percent, divide the part by the whole, and then multiply by 100.

$$\frac{Part}{Whole} \times 100 = Percent$$

For example, suppose 13 green candies are in a bag, and the total number of candies in that bag is 58.

$$\frac{13}{58} = .224 \times 100 = 22.4\%$$

Find the percentage of each colored candy to the total number of candies in the bag. Do this for each of the three sample bags, and write this figure on your table.

- Sample Size—Scientists often use the concept of *sample size* to learn what a large group might be like based on data gathered from a small group. The size of the sample group should be big enough to give a true picture of the larger group.

  Is the data you gathered by only evaluating three bags accurate enough so you can make statements that are true of every bag of M&M's candy?

  If you found that the number of each colored candy is very different in each of the three bags, would you then predict that when you open a fourth bag, the number of colored candies will also be random in that bag?

  What was the total number of candies in each bag? Was it always the same? Is there a range? For example, if one bag had 55, one 58, and one 57, then the range of candies is from 55 to 58. By your examination of three bags of candy, would you predict that every bag you open in the future will not contain exactly the same number of candies?

- Weight Comparison—On the label of the candy bag is the weight of its contents. In the case of an M&M's bag, 47.9 grams. But, did you find each bag had a different number of candies? What can this mean? Either the weight shown on the label is only an approximate or minimum weight, or each candy does not weigh exactly the same!

  Set several candies next to each other. Do they all look the same size or are some bigger than others?

  If you have access to a very accurate gram weight scale at school, weigh two bags and compare their weight.

You can also build a simple balance beam to compare the weight of two bags. Use a coat hanger suspended from a door jam with a candy bag hanging from each end to see if they are in balance (the hanger will tilt if one bag is heavier than the other).

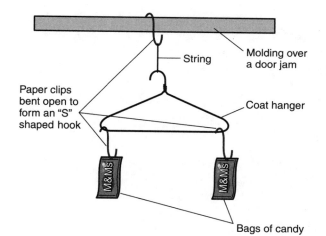

String
Molding over a door jam
Paper clips bent open to form an "S" shaped hook
Coat hanger
Bags of candy

Results

Write down the results of your experiment. Document the data collected.

Conclusion

Come to a conclusion as to whether or not your hypothesis was correct.

Something More

Determine unit cost. If a bag of candy costs 60 cents and it contains 55 candies, how much does each individual candy cost? This is called *unit cost*. Divide the total cost (60 cents) by the total number of candies (55):

$$\frac{.60}{55} = 0.0109 \text{ or } 1.09 \text{ cents each}$$

How much does each candy cost in a bag that contains 58 candies? Did you get more candies for your money?

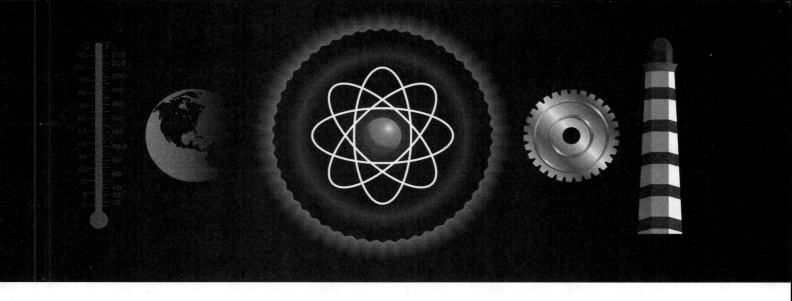

Choices

Behavior: The position of an item will determine the selection by handedness (left hand/right hand) over color

Suggested Entry Category

- Behavioral & Social

Purpose or Problem

The purpose is to determine which is the stronger trait for selecting objects: the location of the object closest to the hand favored (right-handed or left-handed) or the association of color, specifically, gender-associated colors (pink and blue).

Overview

"Pink is for girls and blue is for boys" is the age-old saying. If two cupcakes are offered to a male, both being identical except that one has pink stripes across the icing and one has blue, will he chose the blue one or will he chose the one that is closest to the hand he favors?

In this behavioral experiment, we try to determine which is the stronger trait among males, handedness or color. Determine if most males will choose a cupcake that is easier to take because of handedness (right handed/left handed) or more inconvenient to take, but is colored blue rather than pink.

If a male is told that two cupcakes are identical except for the color of the icing stripes, do you think the color will influence the decision as to which one he will choose? Because most of us have heard the saying "Pink is for girls and blue is for boys," do you think a boy would unconsciously select a blue-striped cupcake over a pink one? Do you think he would choose a blue one out of fear that someone watching might ridicule him for taking a pink one because he thinks pink is for girls?

Or, do you think convenience has a stronger influence in his selection? When two cupcakes are placed in front of a person, a right-handed person may tend to take the cupcake on the right, because it is closer and easier to take, while a left-handed person may go for the one on the left.

Hypothesis

Hypothesize that more males will select a blue-striped cupcake over a pink-striped cupcake, even though the pink-striped cupcake will be closer to the hand he favors (right handed/left handed).

Materials' List

- Two dozen vanilla cupcakes with white icing
- Food coloring
- Two small dishes (cereal bowls) to mix food coloring
- Spoon
- Eyedropper
- Between 10 and 20 male friends
- Small serving tray
- Paper and pencil
- Camcorder or camera and video tape recorder (optional)
- Possible adult supervision needed

Procedure

Bake or purchase a batch of vanilla cupcakes with white icing. The size and shape must be held constant.

Using food coloring, pour a small dish of blue (a standard food color) and mix a dish of pink (stirring several colors together).

With an eyedropper, draw blue parallel lines on half the cupcakes. Draw pink parallel lines on the rest of the cupcakes. You may want to mix a batch of icing and mold it into strips, coloring some blue and some pink. Then, lay the strips in parallel across the top of the cupcakes.

The variable in this experiment is the color of the stripes on the cupcakes. The blue-colored cupcakes will constantly be

Project 19: Choices

positioned on a serving tray, so they are on the opposite side of the person's handedness.

The object of the stripes is to give enough color to the top of each cupcake to establish a definite color difference between them. We do not want to make all the icing solid blue or solid pink, because people tend to associate color with taste. It is important that the test subjects do not make a choice of cupcakes based on how they think a cupcake will taste. Simply being told that all the cupcakes are identical except for the color stripes on the icing may not be enough to unconsciously convince them.

Next, gather a dozen or more male test subjects. We must know if each subject is left-handed or right-handed, so we will ask them. But, we do not want to tip them off that this experiment has something to do with their handedness. So, disguise the question by asking five or six nonrelated questions, surrounding the handedness questions by other questions. When polling each person, write down all their answers, so they do not suspect you are only interested in the data about their handedness. You may choose to use a computer to type the questions and print copies for the subjects to fill in themselves.

Suggested questions:

1. What is your favorite music group?
2. Are you right- or left-handed?
3. Is your bedroom on a first or second floor?
4. Name a TV show you try to never miss.
5. What time do you usually go to bed at night?

Obviously, question two, which is buried within the group of questions, is the only one we care about, but it is camouflaged with other questions.

Once you know if a person is right- or left-handed, place a blue cupcake on a tray that will be on the opposite side of her handedness and position a pink cupcake on the other side. For example, if a person is right-handed, place a pink cupcake on the side of the serving tray that will be nearest her right hand when the tray is presented to her and place the blue cupcake on the left side. She will have to extend her arm further to reach the blue cupcake than the pink one.

This experiment lends itself well to video tape recording for later evaluation and to enhance a presentation in a science fair.

Did more boys pick blue cupcakes or did they pick the handier pink cupcakes?

Be sure none of your test subjects has an allergy or food-related problem with cupcakes before they eat one.

Results

Write down the results of your experiment.
Document all observations and data
collected.

Conclusion

Come to a conclusion as to whether or not
your hypothesis was correct.

Something More

1. Repeat the experiment testing girls
   instead of boys. Place the pink
   cupcakes further out of reach and the
   blue ones closest to their handedness.

2. Are there shape preferences by
   gender? Use cookies and cookie-cutter
   molds to make stars, triangles, hearts,
   and circles. Do males tend to select
   different kinds of shapes than females
   do?

Plants Exhale

A plant produces more oxygen when light intensity is increased

Suggested Entry Category

- Botany

Purpose or Problem

The purpose is to determine if the intensity of light on a plant's leaves affects the plant's oxygen output.

Overview

Chlorophyll (a chemical found in the green leaves of plants) and light collected by the leaves of plants combine with water and carbon dioxide to start a process that causes the leaves to release oxygen.

How critical is light to this process? If less light reduces the amount of oxygen produced, you might conclude that plants living underneath a thick canopy layer in a forest do not produce as much oxygen as those in direct sunlight.

Hypothesis

Hypothesize that *elodea*, a common aquarium plant, will produce more oxygen when subjected to a higher intensity of light.

Materials' List

- Two small aquarium tanks
- Two test tubes
- Two elodea aquarium plants (available at pet shops)
- Two table lamps
- One 15-watt light bulb
- One 100-watt light bulb
- Two dark areas (for example, closets)
- Possible adult supervision needed

Procedure

Fill two small aquariums (or very large, tall, clear glass jars) with water. Place an elodea plant in each one. Elodea is a common underwater plant that many people who have pet fish use in their aquariums.

Fill a small test tube with water. Place your finger tightly over the top to prevent the water from escaping and air from entering. Turn the test tube upside down and lower it into the water, capturing as much foliage from the elodea plant inside the tube as possible. All the plant's leaves do not have to fit inside the tube. Also, it is not necessary for the test tube to remain completely

vertical. It can tilt a little if it needs to rest against the side of the tank for support.

Similarly, place a water-filled test tube over elodea foliage in the second aquarium. The elodea, aquariums, and size of the test tubes are constant. The variable in this project is the amount of light each plant receives.

Place each aquarium in a dark area of your house, for example, a closet or an unused room with the drapes closed.

Find two table lamps, and remove the lamp shades. Position one lamp next to each aquarium. In one lamp, place a 15-watt light bulb; use a 100-watt bulb in the second lamp.

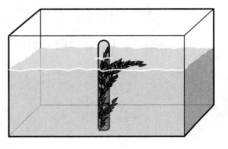 

15 Watt bulb

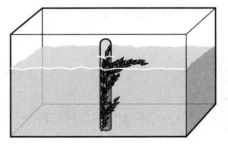 

100 Watt bulb

Be sure the electric cord is not plugged into the wall when you are changing light bulbs. Also, make certain the lamps are not touching any curtains, drapes, or any paper or cloth material that could pose a fire hazard. Keep the lamps alongside the

aquariums. **Do not attempt to put them on the top, where they could come in contact with the water and create an electrical hazard.**

Make daily observations to see if any oxygen is accumulating in the test tubes.

After observing for many days, is the quantity of oxygen different in the test tubes? Which one has accumulated the most?

Results

Write down the results of your experiment. Document all observations and data collected.

Conclusion

Come to a conclusion as to whether or not your hypothesis was correct.

Something More

1. Is there an intensity of light beyond which there is no advantage increasing it?

2. In school science labs, a burning splint is often used to detect the presence of oxygen in a small test tube. A "pop" or tiny explosion occurs if oxygen is present. Use the burning splint technique to confirm that the gas collected from the elodea is oxygen. **Do this under the supervision of a school science instructor with proper safety precautions in place.**

Melting Mountains

Alluvial runoff from melting mountain ice

Suggested Entry Categories

- Environmental Science
- Physics

Purpose or Problem

Mountain material is lost to erosion every spring as snow and ice on mountain peaks melt. We will attempt to determine if the rate at which the ice melts (due to temperature) has an effect on the amount of material lost to erosion.

Overview

Alluvial deposits are the materials (soil, rock, debris, and so forth) that have accumulated from being transported by moving water. Alluvial formations are created at the deltas of rivers, as material is deposited by the currents of water.

Every spring, snow and ice on mountains begin to melt. As gravity and the steepness of mountain sides accelerate the flow of water, materials from the mountain are carried along with the water. If spring temperatures are warmer than normal, will this cause more material to be carried off the mountain than if temperatures were more moderate? If so, mountain erosion caused by melting snow and ice would be in direct relationship to temperature.

Hypothesis

Hypothesize that more alluvial material will be deposited by melting ice when the temperature increases.

Materials' List

- Two plastic gallon milk jugs
- Two Styrofoam trays
- Two 6-inch-long pieces of 2×4 lumber
- Water
- Two cups of very fine sand
- Scissors
- Use of a freezer
- Gram weight scale
- Kitchen measuring cup
- Heat lamp
- Ruler

Procedure

Two model mountains with melting ice will be set up, with one melting at a much more rapid rate. Held constant in the experiment will be the quantity of sand, ice, and the slope of the runoff. The variable will be the temperature.

Find two small Styrofoam trays. Styrofoam trays are often found in grocery stores with ground beef, veal patties, and other meats packed in them. With scissors, cut the lip off one end of each tray.

Weigh one of the Styrofoam trays on a gram weight scale, and then fill a kitchen measuring cup with very fine sand. Pour the sand onto the Styrofoam tray, positioning it at the center of the tray. Weigh the tray with sand. Subtract the weight of the tray (the *tare* weight) to find the weight of the sand.

The second tray must be filled with exactly the same quantity of sand as the first tray. Weigh the second Styrofoam tray. Pour a measuring cup of fine sand onto the center of the tray, and then place it on a scale. Subtract the weight of the tray. Add or take away sand from the tray until the second sand pile weighs the same as on the first tray.

Manipulate the sand on both trays until both piles of sand are about equal in size and shape. Use a ruler to measure the height of each pile.

Carefully, using scissors, cut off the bottom part of two plastic gallon milk or water jugs. Cut about 1½ inches up from the bottom, leaving small plastic bowl-like containers. Fill each jug bottom with water to a depth of 1 inch. Place in a freezer.

When completely frozen, remove the two 1-inch slab ice cubes from the plastic jug bowls.

Use a piece of 2×4 wood (or several books stacked on top of each other) to tilt the tray with sand. Position the cut-away end of the tray at the bottom of the slope, and rest it in one of the plastic jug bowls. This will

Results

Write down the results of your experiment. Document all observations and data collected.

Conclusion

Come to a conclusion as to whether or not your hypothesis was correct.

capture any runoff water and sand that is carried down. Carefully set an ice slab on top of each sand pile. This will create a model of melting ice on top of a mountain.

Place the whole device in a cool location somewhere in your home.

Make a second device with the other tray, sand, ice, 2×4, and bowl. Set this one in a place where a heat lamp can be safely positioned over the top of the ice. You can use an infrared bulb, available at pharmacies or possibly from your school science teacher, as a heat lamp.

When the ice on both trays melts completely, remove the two plastic jug bowls. Let the water collected in them evaporate. Then, using a gram weight scale, weigh the remaining particles of sand in each bowl. Was more sand deposited from the heat lamp mountain?

Something More

1. Repeat the previous experiment, but this time, use coarse particles of sand. Hypothesize that not as much material will be deposited because of the bigger size of the particles.

2. How does *pitch* (the slope of the runoff) affect erosion?

Sounds Fishy

Determining if goldfish have water temperature preferences

Suggested Entry Categories

- Behavioral & Social
- Zoology

Purpose or Problem

Find the relative temperature (warmer or cooler) that pet goldfish prefer.

Overview

Pet goldfish can withstand a rather wide range of temperatures. Are they more comfortable in water that is above or below room temperature? Finding out which temperature they prefer, and then giving them that temperature, may make for healthier, longer-living pets.

Hypothesis

Hypothesize that water temperatures above room temperature (within reason) are more desirable to goldfish than water temperatures below room temperature, as evidenced by their spending more time in the area of an aquarium they prefer. (Or, you may choose to hypothesize the opposite.)

Materials' List

- Piece of Plexiglas
- Nontoxic bonding glue for plastic and glass
- Aquarium
- Several goldfish
- Two thermometers
- Fish tank heater
- Several ice cubes

Procedure

Position a board or a piece of Plexiglas in an aquarium tank. It should extend from the top to the bottom of the aquarium, and when placed in the center of the tank, a 1-inch gap should be around one of the sides to allow the fish to freely swim from one side of the tank to the other. Secure with glue and let dry.

Get two thermometers. Lay them side-by-side on a table and allow them to sit for about 15 minutes to stabilize at room temperature. Both thermometers should read the same temperature. If one reads differently, mark it with a marker and note the difference in temperature. You will need to account for this difference when you make temperature readings in the water.

Fill the aquarium with water. Let the water stand for several hours until it reaches room temperature.

Place a fish-tank heater in one side of the aquarium. Place several ice cubes in the other side. Position two thermometers in the tank, one in each side.

Introduce the goldfish into the water and observe them for several hours. Do the fish spend more time in one section than the other? What is the temperature difference between the two sides?

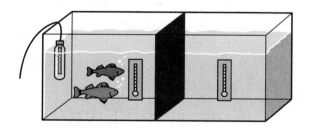

Results

Write down the results of your experiment. Document all observations and data collected.

Conclusion

Come to a conclusion as to whether or not your hypothesis was correct.

Something More

1. Can you entice the fish to go into the side of the aquarium that has a less-desirable water temperature by always feeding them on that side?

2. Can goldfish be trained? Make a tapping sound on the side of the aquarium every time you feed them. After several weeks, will they come to the feeding area when they hear the tapping, even if no food is present?

3. Do goldfish prefer light or dark, or do they have no preference for either? Make one partition of the aquarium well lit and the other darker by covering the sides with dark construction paper.

4. Are there other factors that might be at play, such as light? Will the result be the same if you turn the tank 180 degrees?

Parallelogram Prevention

Simple bracing can greatly increase a structure's capability to maintain its shape under stress

Suggested Entry Categories

- Engineering
- Physics

Purpose or Problem

The purpose is to find the best placement of braces to strengthen a square wooden frame.

Overview

The science discipline of engineering has many branches: chemical, electrical, aeronautical, biochemical, nuclear, and mechanical.

Mechanical engineering requires knowledge of the behavior of materials and their physical properties to design buildings, machines, aircraft, bridges, skyscrapers, overpasses, dams, and other structures. Mathematics, physics, experimentation, and testing are important aspects of mechanical engineering.

Physical structures that undergo stress are often given extra support by using braces at key positions on the structure. Amazingly, braces sometimes don't need to be very strong to be effective. Their placement in the structure is more critical than their strength. Yet, they can give tremendous strength to the structure they are bracing.

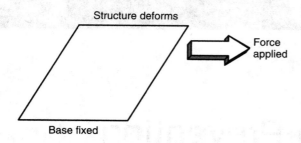

Structure deforms

Force applied

Base fixed

This project requires wood to be cut to specific lengths and holes to be drilled in the wood. **If power tools are to be used, make sure you have adult supervision. Observe all safety precautions using power tools.**

Hypothesis

Hypothesize which of three different bracing forms can best help to keep a square wooden structure true to form when an unbalanced force is applied.

Materials' List

- One 5-foot piece of rope
- Pulley
- Two 2-inch-long bolts with nuts
- Several plastic gallon milk/water jugs

- Four large screw hooks
- Lumber: five 8-foot-long pieces of ½-inch pine of any width (1 to 4 inches)
- Lumber: several 3-foot-long shims
- Lumber: 2-foot piece of 2×4
- Several dozen nails about 1½ inches long
- Several buckets of sand
- Towel
- Scale
- Wood saw
- Wood drill
- Hammer
- Use of a table
- Adult supervision, if using power tools

Procedure

Make four 2-foot-square wooden frames using ½-inch pine boards. The boards can be of any width, from 1 inch to 3 or 4 inches. Use only one nail at each corner.

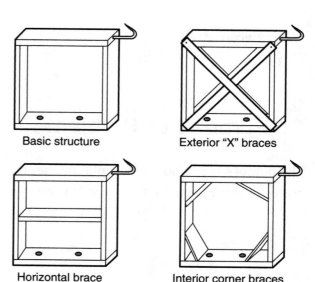

Basic structure

Exterior "X" braces

Horizontal brace

Interior corner braces

Stand the frames vertically, as shown. Drill two holes in the bottom of each frame. The holes must be placed 12 inches apart.

Put one screw hook into the end of each frame at the top of one side, as shown above.

Take one of the frames and add exterior *X* braces from corner-to-corner on both sides. These braces need not be made out of strong wood. Thin wood, such as shims or wood used for making latticework, will suffice.

Take another frame and cut a horizontal brace out of the pine wood. Use one nail through each side to mount it horizontally across the interior center.

Using the pine wood, cut four 4-inch pieces and nail them into the inside corners as shown.

This completes the four structures to be tested. Next, construct a jig, so each structure can be tested.

Cut a 3-foot-square piece of ½ or thicker plywood. At one edge, securely mount a 2-foot vertical piece of 2×4 wood. You may need to strengthen it by using a large metal elbow angle brace behind it. Mount a pulley at the top of the board.

Near the middle of the plywood, drill two holes spaced 12 inches apart. The holes must be in line behind the 2×4 piece, as shown. Push a 2- or 3-inch bolt up through each hole. This completes our test jig.

Position the basic wooden structure on the test jig, so the bolts come up through the holes in the structure. The hook should be on the side facing the pulley. Screw the nuts down onto the bolts and tighten. The structure is now secured to the jig.

Set the jig on a table. Place a towel or something soft underneath the jig to be sure the heads of the bolts do not scratch the surface of the table.

Tie a piece of rope onto the hook. Drape it over the pulley and let it hang down. Tie the other end onto several empty plastic gallon milk or water jugs, so the jugs hang suspended and do not touch the floor.

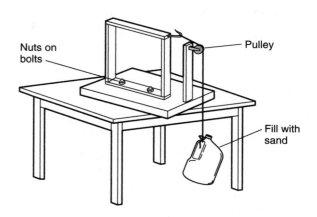

Begin adding sand to the jugs until the square wooden structure begins to show signs of skewing. If the test jig begins to tip, place some heavy weights on the plywood base or have a friend sit on it! Weigh the jugs of sand.

Repeat the procedure for each of the three structures that have braces added. Record the weights.

Were all the brace structures able to sustain more lateral force? Which brace method worked best? Which was the least beneficial?

Results

Write down the results of your experiment. Document all observations and data collected.

Conclusion

Come to a conclusion as to whether or not your hypothesis was correct.

Something More

Compare the strength of a simple wooden square structure where the top and bottom lengths of wood are equal, and the two shorter side pieces are equal, to a structure where all sides are equal, with each corner overlapping the next, as shown in this illustration.

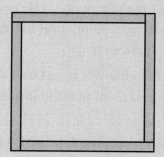

A Taste of Plant Acid

Determining if a vegetable has a more acrid taste if it has a higher pH

Suggested Entry Categories

- Botany
- Chemistry

Purpose or Problem

The purpose is to determine if pH could be a factor in giving an onion an acrid taste.

Overview

Some vegetables have a strong, sharp, "acrid" taste, sometimes even stinging your tongue and nose. A good example is an onion. Yet, even among the onion family, there are differences in the strongness of their taste. Does the *pH* (the measure of alkalinity or acidity) differ in different types of onions (American onion, Spanish onion, and so forth)? If one type of onion has stronger taste than another, do you think the pH of the stronger onion is more acidic?

One characteristic of chemical substances is the amount of acid or base they contain. Foods that contain weak acids taste sour, for example, lemon or lime juice, and pickles. Strong acids may be hazardous, as they can damage your skin and other parts of your body. The opposite of an acid is a *base*, also called an *alkali*. Bases have a slippery feel to the touch. They taste bitter. Examples of substances that are bases include ammonia and other cleaning products, milk of magnesia, baking soda, and soap. Products that unclog household drains contain strong bases, and they can be hazardous to touch.

In the same way as a ruler is used to measure the length of an object, and a thermometer is used to show how hot something is, chemists have created a scale to measure how much acid or base a substance contains. This measurement tool is called the *pH scale*. Technically, the term "pH" means "the potential of electricity for positive hydrogen ions," because chemists can use electricity to do the measurement.

The pH scale goes from 0 to 14, 0 as the strongest acid, 7 s neutral (in the middle), and 14 as the strongest base. Pure water has a pH of 7. If you have a swimming pool, you may have used a pH water-test kit, where a sample of water is collected, and a few drops of a special chemical are added and mixed with the water. The resulting color of the water is matched against a color comparator chart to find the exact pH.

One way to measure pH is by using litmus paper. *Litmus paper* comes in different colors to measure different ranges of pH. *Red litmus paper* turns blue in a base solution. *Blue litmus paper* turns red in an acid solution. A color chart is used to compare the color litmus paper turns to a pH number.

The pH Scale

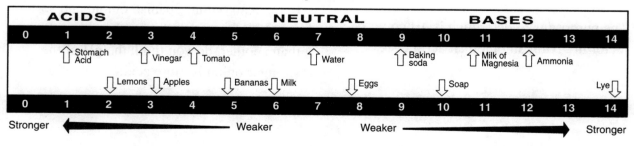

Hypothesis

Hypothesize that among several types of onions, the ones with the stronger (more acrid) taste have lower pH values (will be more acidic) than the others.

Materials' List

- As many different types of onions as you can find
- pH test kit with color comparator

Procedure

Gather as many different types of onions as you can. Squeeze juice from each type and use a pH test kit to determine the pH of the juice. Taste a small amount of the juice of every onion, and then compare the strongness of each. Compare the tastes to the pH of each onion. Do the onions with a stronger taste have a lower pH?

Note: If the pH is more acidic in stronger-tasting onions, this poses an interesting correlation. However, it does not prove that pH is the only factor involved in making the onions taste more acrid. Additional research would have to be done.

Results

Write down the results of your experiment. Document all observations and data collected.

Conclusion

Come to a conclusion as to whether or not your hypothesis was correct.

Something More

Compare the pH of various other vegetables. Is there an association with sharp taste to pH? Does a red bell pepper have a different pH than a green bell pepper? Compare the pH of various citrus fruits to their taste.

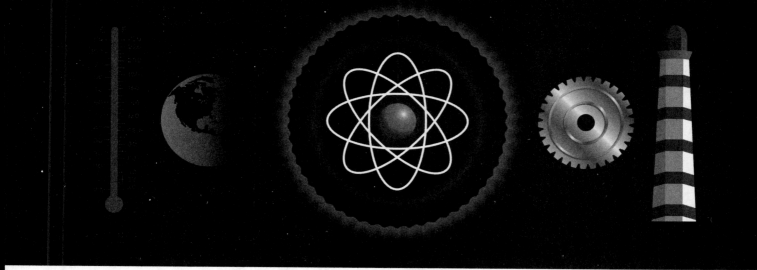

Project 25

Split and Dip

Testing a strategy for making money in the stock market

Suggested Entry Category

- Math & Computers

Purpose or Problem

The purpose is to determine if a specific strategy for buying or selling stock in the stock market might be profitable over a period of time.

Overview

What's the secret to making money in the stock market? It's simple: buy low, sell high. That's it! Everything else is just "noise." Buy a stock when its price is low and sell it at a later date when the price has appreciated. Of course, the hard part is picking a stock whose price will appreciate.

Most successful investors buy stock in a company they have thoroughly researched and they believe the company has great potential for increasing its earnings in the future. When a company has increasing earnings, the price of its stock usually increases. Buying stock in a good company and holding onto it for many years is a great way to build a nice nest egg.

But, some people try to get a much quicker return on their money. They attempt to buy and sell stocks in a very short period of time, perhaps over a period of several weeks or months. Some even buy and sell within the same day; they are known as *day traders*.

Short-term investing can be very risky, however, and, many times, an investor will lose money or miss out on really big gains because a stock was sold prematurely. Patience usually gives investors an edge.

Nevertheless, the fearless investor is lured to a quick profit by short-term trading, just as a gambler is drawn to the chance for a big win with the pull of a slot-machine handle.

Stock market trading, however, is not quite the same gamble as pulling the handle on a slot machine. Naturally, some luck is involved and some uncontrollable factors are at play. But the odds are more in your favor in the stock market, where you can do your homework researching a company and its products, and developing strategies based on the past performance of a stock.

Stock market gurus are everywhere, selling newsletters and services, each one making recommendations as to which stock to buy, based on their "unique" strategies. How accurate are the predictions by these gurus? Their theories must be tested over time with a large sample of stocks.

Let's propose a strategy of our own, and then, over a period of two months, test our theory with a large sample size. Our strategy will be a short-term play. We will attempt to buy low and sell high within a few weeks or months to make a small profit.

Our strategy is to buy a stock if it takes a slight "dip" within the first three weeks of doing a stock split, and then selling the stock several weeks after that as the stock begins to climb.

What is a stock split? A *stock split* is when a company divides its total number of shares of stock to make more shares available. For example, suppose a stock is selling for $100 a share and you own 20 shares. If the company does a 2-for-1 split, you will get two shares of stock for every one share you have, but the price per share will be cut in half. After the stock split, then, you will own 40 shares of stock valued at $50 per share. Think of a 2-for-1 stock split as somebody giving you two five-dollar bills and you giving them a ten-dollar bill. You end up with two bills, but the total dollar value of the money you have has not changed. You got two $5s for a $10.

Companies may chose other ratios for splitting, for example, 3-for-1 or 3-for-2.

Why do companies split their stock? Splitting makes the price of their stock cheaper, so it is more affordable. They hope people will then buy more of it, which will cause the stock's value to increase.

Why would a stock make a temporary dip in price following a split? Once a stock split is announced, traders often start buying shares, which drives the price up a few weeks before the stock is scheduled to split. Once the split takes place, traders begin to sell their shares to lock in profits from the presplit price run-up. Historically, stock splits were a nonevent. A split is nothing more than two $5s for a $10. But, in recent years, a split has been perceived as a positive move, because stocks that announce splits have usually been very good stocks. So, the stock often goes up when a split is announced because people buy into what they think is a positive move. Once the stock splits, some traders sell their stock to capture a small profit from the stock price rising between the time of the announcement and the actual split date. That selling may cause the stock price to take a slight drop. This often happens within the first three weeks following the split. After the dip hits a bottom, the stock price may begin to rise again. There is a tendency for the stock to regain lost ground and head up toward its previous high (which may take a year or two) if it is a good company, which is why the stock went up so high in the first place.

Buying a stock on the slight dip after a split is an opportunity to buy low, and when the stock begins to climb shortly after that, to sell high.

Hypothesis

Hypothesize that you will make a profit the majority of the time by buying stocks that take a slight price dip within the first few days or weeks following the split, and then selling the stock within a month or two as the price moves upward. We will only buy stocks that take a dip within the first three weeks following a split. We will sell the stock within two months following the split. Stocks will be bought and sold *on paper* (not using

real money). After several months, our wins and losses will be tallied. Our hypothesis is that we will make an overall profit, even though there will occasionally be some losing stock plays.

Materials' List

- Computer with a modem and an Internet connection
- Computer printer
- Financial daily newspaper
- Two to three months of time
- Calculator or computer spreadsheet program (optional)

Procedure

Research to compile a list of stocks that have announced they are going to split. Companies that announce stock splits are listed in daily financial newspapers (*The Wall Street Journal*, *Investors Business Daily*, and so forth) and announced on financial news programs on TV and on financial channels (CNBC). Many stock market sites on the Internet also list upcoming stock splits. Some sites include:

- http://biz.yahoo.com/c/s.html
- www.coveredcalls.com

You need not wait for splits to take place. You can find out when stocks have split in the past by looking at their stock charts (which indicate when a stock split) or by checking these Internet sites:

- http://quote.yahoo.com
- www.briefing.com
- www.cboe.com
- www.freerealtime.com
- www.stocktools.com
- www.theonlineinvestor.com

This project requires viewing a lot of stock charts to observe the price behavior patterns following stock splits. The Internet has hundreds of free financial sites where you can get stock charts. Some of the best include:

- http://quote.yahoo.com
- www.bigcharts.com

Stock charts are displayed by entering a company's *ticker symbol*, the symbol used by the stock exchanges to represent each company. For example, Disney is DIS, Coca-Cola is KO, Ford Motor Company is F. If you do not know a company's ticker symbol, you can usually obtain it from the Internet site where you are getting your stock charts. The daily financial papers also list all companies alphabetically, along with their ticker symbol.

The Internet is a fast-changing medium, and the sites we have just listed for research and charting may or may not still offer these free services. If that happens, use an Internet search engine to search for stock market quotes or stock splits.

Once you compile a list of stocks that are due to split (at least a dozen stocks split every month), write down the name of each company, the ticker symbols, and the date the stock is due to split. Once a day, check a financial Internet site or a financial newspaper and write down the price of your stock. You may want to do this every morning or evening.

After a stock splits, watch for a drop in price. When you find a stock that is in a downward trend for several days, watch it until it appears to level off or head back up. At that time, do a paper trade, that is, pretend to buy 100 shares of the stock. Write down the cost per share and how much money you spent. Timing the market is very difficult. You may think you are buying the stock at the lowest part of the dip, only to later discover that the stock dips even more. Nevertheless, hold on to your stock, even if you think you didn't buy it at its lowest point.

Once you purchase a stock, continue to write down its closing price each day. Watch it for one or two months. If the stock rises to a point where you feel you would make a nice profit if you sold it, sell your shares. To arrive at your profit, subtract your cost from the money you made selling your shares. Determine the percent return on your money by dividing the profit by the cost, and then multiplying the answer by 100 to show percent. For example, suppose you bought 100 shares of Kmart stock at $12 per share, for a cost of $1,200. You sold the stock at $13½, or $1,350. The profit is $150. $150 divided by $1,200 equals 0.125, times 100 to show percent is 12½ percent. That's a very good return over two months!

Our rule is to only buy stocks that show a dip pattern within three weeks of a stock split. If a stock does not take a dip, do not use it in the project.

Another rule is to sell any stock you buy on dip that continues dropping more than 20 percent. We will take the loss, but we will preserve our money, so we don't lose too much. Not all stocks do well after a stock split, as evidenced by Disney (DIS) in 1999.

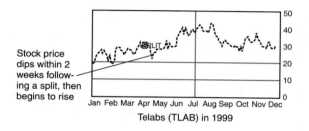

Stock price dips within 2 weeks following a split, then begins to rise

Telabs (TLAB) in 1999

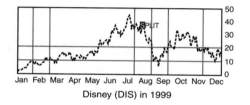

Disney (DIS) in 1999

After several months of tracking stocks, buying, and selling, declare a stop to your project. Total your wins and losses. Did you make more money than you lost? Do you think this is a strategy that could be profitable if real money were used to do the trades?

Results

Write down the results of your experiment. Document all observations and data collected.

Conclusion

Come to a conclusion as to whether or not your hypothesis was correct.

Something More

1. Continue to follow the stocks for an additional two months. Would more profit be made by holding the stocks longer?

2. Research news on any stock that drops after a split and continues to drop for a long time. Is there a reason why the stock price is being beaten down?

Johnny Applesauce

Cinnamon: A mold inhibitor

Suggested Entry Category

- Microbiology

Purpose or Problem

The purpose is to reduce spoilage of applesauce from microorganisms (mold) by adding cinnamon to it. The cinnamon would then serve a dual purpose, acting as both a flavor enhancer and a mold inhibitor.

Overview

For centuries, people have used spices to season their foods. Many kinds of commonly used spices exist, including allspice, cinnamon, cloves, ginger, mustard, nutmeg, and pepper. These spices come from a group of plants.

Early cultures also realized some spices had properties that helped preserve foods. The storage of food was very important to early civilizations. They did not have the sophisticated techniques we have today to prolong the freshness of foods, which include chemical additives, refrigeration, and vacuum packing.

Microorganism growth can quickly shorten the time in which a food must be consumed. Because many people like to sprinkle cinnamon on their applesauce, they would not mind the addition of a natural, complementary spice like cinnamon being used on their applesauce as a preservative. If your project reveals that cinnamon is a good mold inhibitor, then would it be a good idea to add a little cinnamon to an opened jar of applesauce that is placed in the refrigerator to eat later in the week?

Hypothesis

Hypothesize that cinnamon is an effective mold inhibitor for applesauce.

Materials' List

- Four plastic 6- or 8-ounce clear plastic drinking cups
- Small jar of applesauce
- Cinnamon
- Tablespoon measure
- ¼ teaspoon measure
- Felt tip marker
- Use of a closet or another dark area at room temperature
- One week of time

Procedure

Place five level teaspoons of applesauce in each of the four small clear plastic cups. With a felt tip marker, sequentially number each cup: 1, 2, 3, and 4.

Cup 1 will be the control cup. Nothing will be added to it.

In cup 2, lightly and evenly sprinkle ¼ teaspoon of cinnamon on top of the applesauce, completely covering its surface.

In cup 3, add ¼ teaspoon of cinnamon to the applesauce and mix thoroughly. This is our "lightly added" mixture.

In cup 4, add ½ teaspoon of cinnamon to the applesauce and mix thoroughly. This is our "more heavily added" mixture.

Set the cups in a dark area at a normal room temperature. Make sure the cups are out of the way, where they will not be disturbed.

After one week, examine each of the cups. Has mold formed in any of the cups? Are the lightly covered or lightly mixed applesauce cups free of mold?

When your project is completed, dispose of the cups in the garbage. **Do not to eat the applesauce from any of the cups.**

Results

Write down the results of your experiment. Document all observations and data collected.

Conclusion

Come to a conclusion as to whether or not your hypothesis was correct.

Something More

1. What is the smallest amount of cinnamon that can be added to the applesauce to make an observable difference in mold growth? We do not want to have to add so much cinnamon that no one would ever want to eat it.

2. If mold did not grow on the covered applesauce, did it not grow because of the cinnamon or simply because it was covered and kept from exposure to the air? Try covering it with other substances, such as flour, and see if mold forms.

3. Determine if other popular spices, such as nutmeg, demonstrate mold inhibitor properties.

Backfield in Motion

The effect of an electromagnetic field on single-celled organisms

Suggested Entry Category

- Microbiology

Purpose or Problem

The purpose is to determine if there is any effect in the behavior (the movement) of common one-celled organisms when in the presence of a static electromagnetic field.

Overview

There has been a great debate over the years as to whether or not magnetism has an effect on living cells. People on one side of the debate believe that living under high-power electric lines causes harmful electromagnetic radiation that damages farmers' crops and is unhealthy for their bodies. The magnetic field created by alternating current traveling through electric power lines is a moving field; the field expands and collapses many times each second. On the other side of the debate are those people who believe that static fields of magnetism (unlike those generated by high-power electric lines) may be beneficial to health. They place magnets on parts of their body to aid in healing various ailments and injuries.

No doubt, you have at some time placed a magnet underneath a piece of paper or cardboard, sprinkled iron filings on top, and tapped it to see a pattern as the filings form lines, showing the otherwise invisible magnetic lines of force.

Bar magnet placed under cardboard

Iron filings arrange themselves to the lines of magnetic force

Magnetism is a phenomenon that is intimately related to electricity. Early scientists realized this when they placed a compass next to a wire through which electricity was flowing. The compass needle was deflected when electricity was flowing in the wire.

When direct current (DC), which is the type of current flow that comes from a battery, travels through a wire, a magnetic field extends out from the wire. Do common

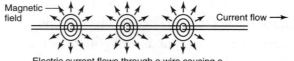

Magnetic field → Current flow →

Electric current flows through a wire causing a magnetic field to surround the wire

single-celled organisms respond to this field by moving toward or away from the wire?

Hypothesis

Hypothesize that several common one-celled organisms will not show any response (to move away from or to move toward the source) to a static electromagnetic field.

Materials' List

- Model train DC transformer
- Small 12-volt light bulb and socket
- Hook-up wire
- Wire cutters
- Small slotted screwdriver
- 40x or 50x microscope
- Microscope slide
- Petroleum jelly
- Adhesive or masking tape
- Live single-celled organisms (euglena, paramecium, flagellum, amoeba, and so forth)

Procedure

Using hook-up wire, connect a model train DC transformer to a 12-volt light bulb to make a complete circuit. Any variable power supply or DC power transformer is appropriate, including one designed for model HO racing car sets. One piece of wire should be very long, so it can be draped across a slide on a microscope. The train power supply has a variable control on it, so the amount of electricity that flows in the circuit can be changed from zero to its full potential.

Position a microscope slide under the microscope and lay the wire across the slide. Use adhesive or masking tape to secure the wire in place. Be sure the wire is flat against the slide.

In the middle of the slide, squeeze a tube of petroleum jelly to form a "donut-shaped" circle. This will act as a wall to contain a tiny liquid pool of microorganisms. If your petroleum jelly is in a jar, use an ice-pop stick or a toothpick to form the petroleum dam.

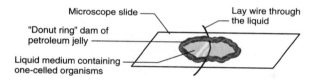

From a science supply house or your high school biology teacher, obtain live single-celled microorganisms. They must be active organisms that are able to move on their own. These include euglena, paramecium, flagellum, and amoeba. Place a few small drops of the liquid medium containing the organisms in the "petroleum pool." The petroleum will not harm the organisms.

Set the microscope with a large-enough field of view so the wire and organisms around it can be seen.

Observe the organisms to be sure they are alive and moving around. Turn on the power supply and watch to see if any response occurs by the organisms to the wire that now has an electromagnetic field surrounding it. Try varying the amount of electricity supplied through the wire. The light bulb will give a visual indication that current is flowing, and its dimness or brightness is also a relative gauge indicating the amount of current.

Results

Write down the results of your experiment. Document all observations and data collected.

Conclusion

Come to a conclusion as to whether or not your hypothesis was correct.

Something More

1. Electromagnetic fields are also present in a wire carrying alternating current (AC). With AC, the electromagnetic field builds and collapses many times each second. Repeat the experiment using AC rather than DC. An AC power supply can be obtained from your local electronics store. An AC doorbell transformer, available at your local hardware store, will also work.

2. Increase the magnetic field by exposing single-celled organisms to the strong alternating magnetic field from a bulk tape eraser (available at electronic stores) used to erase audio and video tapes. Observe the organisms.

Green No More

Concepts in chlorophyll

Suggested Entry Categories

- Botany
- Environmental Science

Purpose or Problem

The purpose of this project is to determine if chlorophyll is present in a leaf that is normally green, but has turned brown.

Overview

In the fall, *deciduous* trees lose their leaves. This is a brilliant display of gold, red, yellow, and orange put on by Mother Nature. During the spring and summer, these leaves are green.

We know that plants use a process called *photosynthesis* to make food. They use chlorophyll, water, minerals, carbon dioxide, and, most importantly, sunlight in the photosynthesis process. From experience, you have no doubt observed that photosynthesis stops when sunlight is cut off from a plant.

Looking under a board that was thrown on a patch of grass reveals the grass has lost its green color.

When a leaf falls from a tree and turns brown, is chlorophyll still present?

Hypothesis

Hypothesize that even though photosynthesis has ceased, some chlorophyll still remains in a brown leaf.

Materials' List

- Alcohol
- Two test tubes
- Two test-tube holders
- One house plant
- Use of a stove burner
- Small cooking pot
- Water
- Scissors
- String
- Adult supervision

Procedure

Chlorophyll can be extracted from a leaf by placing it in a boiling bath of alcohol. If chlorophyll is present, the alcohol will begin to turn green. As a comparator, the more green the color of the alcohol, the greater the quantity of chlorophyll.

Alcohol is highly flammable, and must be handled with care around heat. Because alcohol boils at a lower temperature than water, we can use a double-boiler system that will make it safer to handle for this project.

On a house plant, locate two leaves that are as close as possible to being the same size. Using scissors, cut one leaf off the plant. Let it sit for several days until no green coloring is left in the leaf. To identify the other leaf of the plant, loosely tie a piece of string around its stem.

Gather two test tubes. Fill each one half-full of alcohol. In one test tube, place the brown leaf. Clip the leaf from the live plant (which you previously identified with a string) and place it into the other test tube.

Bring a small cooking pot filled with water to a rapid boil on a stove burner. Turn off the burner.

Using test-tube holders, lower both test tubes into boiling water. Position the holders and tubes, so the test tubes rest upright in the pot, or at least at a steep enough angle so no water gets into the tubes. **An oven mitt may offer additional safety and comfort when you work around the pot.** Let the test tubes remain in the water for 10 to 15 minutes.

Remove the test tubes and observe the color of the alcohol. Is green present in both test tubes? If so, does one test tube contain more green than the other?

Results

Write down the results of your experiment. Document all observations and data collected.

Conclusion

Come to a conclusion as to whether or not your hypothesis was correct.

Something More

1. Is any chlorophyll present in grass that has turned white because of sunlight deprivation?

2. Are vegetables that are green (peas, string beans, lima beans, lettuce, spinach, kale, mint, and so forth) that color because they contain chlorophyll? Check for the presence of chlorophyll in green vegetables. Spinach and mint are leaves. Are your results different for these than for peas and beans?

Not Just Lemonade

Determining if the addition of lemon to cleaning products is strictly for marketing purposes

Suggested Entry Categories

- Behavioral & Social
- Chemistry

Purpose or Problem

The purpose is to determine if the addition of lemon to cleaning products aids in actual cleaning or if the only reason it is an ingredient is to increase sales (because most people associate the scent of lemon with cleanliness).

Overview

Lemons and lemon juice have long been used in cooking, drinks, and candies. But, walk down the aisles of a supermarket today and you'll see many cleaning products that boast of containing lemon (or at least a lemon scent). Look at the advertising on the labels of general-purpose liquid cleaners, shower and tub cleaners, dishwashing soap, glass cleaners, and shampoo. And, you'll see marketing phrases highlighting lemon, such as "Lemon Fresh!" and "New Lemon Scent!"

The scent of lemon is a smell most people psychologically associate with cleanliness and freshness. Manufacturers have capitalized on this fact, as is evidenced by their promotion of a lemon ingredient in their advertising.

Is lemon added to household cleaners only for marketing purposes or does lemon juice actually have cleaning properties?

Hypothesis

Hypothesize that lemon juice is included as an ingredient in many cleaning products, not only for its psychological association with cleanliness and freshness, but also because it has true cleaning properties. (Or, you could hypothesize the opposite, that the addition of lemon is solely for marketing purposes.)

Materials' List

- Several fresh lemons
- Tarnished penny
- Tarnished piece of silverware (fork, spoon, bowl, and so forth)
- Use of a glass window (car or house)
- Use of a kitchen countertop
- Copper bottom pot
- Chrome surface
- Plastic item
- Dirty dinner dish
- Greasy pan bacon was fried in
- Kitchen strainer or small piece of screen
- Cup
- Old rag or washcloth

Procedure

Squeeze the juice from several lemons through a small piece of screen or a kitchen strainer into a cup. The strainer will keep pulp to a minimum.

Evaluate the effect of pure lemon juice as a cleaning agent on various surfaces by rubbing them with an old rag or washcloth. Surfaces to evaluate include a copper penny or a copper bottom pot; a tarnished piece of silverware; a chrome surface, such as a bathroom faucet; a kitchen countertop; a dirty dinner dish; a glass window in your home or a car windshield; and something made of plastic (a child's toy or a telephone). Try to clean as many different surfaces as you can. Do not, however, try to clean any valuable object, such as a wooden dining room table or the upholstery of a living room couch. You do not want to run the risk of staining or damaging any valued object. Lemon juice is acidic. Instead, use an old piece of fabric or an old discarded piece of furniture on which to experiment. Also, when you take a shower, try lemon juice as a shampoo and a soap. *Just be careful not to get it into your eyes.*

Something More

How many products (other than food-related products) can you find that have lemon or lemon scent as an additive? Soap? Laundry detergent? Car air fresheners? Spray starch for ironing?

Results

Write down the results of your experiment. Document the effect of lemon juice on each surface. Did it make the surface cleaner? Did it leave the surface sticky? Did it leave a film on glass? Was it effective at dissolving grease?

Conclusion

Come to a conclusion as to whether or not your hypothesis was correct.

Less Is More

Determining if pH increases as standing rainwater evaporates

Suggested Entry Categories

- Chemistry
- Earth Science
- Environmental Science

Purpose or Problem

The purpose is to determine if rainwater becomes more acidic as it evaporates and concentrates the contents that are left. Does less water make more acidity?

Overview

One environmental concern today centers on the damaging effects of acid rain. *Acid rain* is a term used to describe precipitation (rain, snow, hail, sleet, and fog) that has a low pH. The amount of alkalinity or acidity in a liquid is measured on a scale called the pH scale. The scale goes from zero to 14, with zero being the strongest acids, 7 neutral, and 14 the most alkaline.

Because the Earth's environment has natural sources of sulfur and nitrogen, it is normal for rainwater to be slightly on the acidic side, having a pH of 5. But when the pH drops to 4, scientists consider this acid rain.

Scientists theorize that acid rain is caused by chemicals in the atmosphere, including sulfur dioxide (produced by industries burning oil and coal) and nitrogen oxide (which comes from automobile exhaust). Winds can carry these airborne chemicals thousands of miles. Acid rain has been discovered in many areas around the world.

Acid rain attacks metal and stone structures, and over a period of time can damage them. Most importantly, it can fall into ponds, lakes, and other bodies of water, where it can make the water too acidic for the animal and plant life to survive.

Does the acid in rainwater become even more concentrated as the water evaporates, thus worsening the effects of acid rain?

Hypothesis

Hypothesize that as rainwater evaporates, the concentration of acidity increases (the pH of the remaining water decreases). Or, hypothesize the opposite to be true.

Materials' List

- 6- or 8-ounce jar
- Large funnel
- Rainy day
- Large, shallow dinner dish
- Sunny window
- Litmus paper pH test kit with color comparator chart

Procedure

Using a 6- or 8-ounce jar or a drinking glass, collect about 5 or 6 ounces of rainwater on a rainy day. Place a large funnel in the top to increase the area of rainwater collection. **Do not go outside to collect rainwater when lightning is present!**

If it doesn't rain enough to fill your jar, put a lid on the jar or cover it with a tight-sealing piece of plastic wrap to prevent evaporation. Then, the next time it rains, set out your collection jar and funnel again.

Indoors, place a large, shallow dinner dish in an undisturbed area that, during sunny days, receives a lot of warm sunlight. Pour as much rainwater into the dish as it will hold.

Using a pH test kit, determine the level of acidity on the pH scale. Once a day, test the pH level, and then write down the date and the pH number. Continue testing daily until all the water has evaporated.

Has the water become measurably more acidic as it evaporated?

You may want to enhance your project by also testing the pH of small areas in your neighborhood daily, where rainwater collects: mud puddles, birdbaths, swales, depressions at the base of downspouts, and similar places.

Results

Write down the results of your experiment. Document all observations and data collected.

Conclusion

Come to a conclusion as to whether or not your hypothesis was correct.

Something More

Acid rain can destroy plant and animal life living in bodies of water. Start a log of daily pH measurements of nearby lakes and ponds. In your daily log, include information on rainfall that occurs (the date and the inches of rain).

Natural Fences

Finding natural pesticide substances

Suggested Entry Categories

- Environmental Science
- Medicine & Health
- Zoology

Purpose or Problem

The purpose is to try to find natural substances that will act as a pesticide that can be safely used around the home.

Overview

Insect pests are something everyone must deal with around their home. Sour flies around fruit and ants in the kitchen can be a nuisance. Hardware stores and supermarkets sell chemical pesticides, often in aerosol bottles, to spray in your home. But, some pesticide chemicals have the potential of being hazardous to our health. We want to be careful about using chemicals around the eating areas of our home, in rooms where children play, and in our yards when well water supplies our home.

Are there more natural substances we could use to keep away common household pests, such as flies, sour flies, and ants? Natural substances (lemon juice or tea, for example) would not only be safer, but they also might smell better than a commercial pesticide product.

Hypothesis

Several hypotheses can be developed:

- Hypothesize that lemon juice can (or can't) be used as a natural deterrent for fruit flies.

- Hypothesize that grapefruit juice can (or can't) be used as a natural deterrent for fruit flies.

- Hypothesize that concentrated tea can (or can't) be used as a natural deterrent for fruit flies. (Also develop similar hypotheses about ants.)

Materials' List

- Several lemons
- Several grapefruits
- Tea bags
- Three shallow bowls or dishes
- Drinking glass
- Several bananas
- Tea cup (or mug)
- Boiling water
- Three spray bottles
- Two paper coffee filters
- Funnel
- Microwave oven (or a tea kettle and use of a stove)
- Anthill
- Glass jar with lid

Procedure

Line a funnel with a paper coffee filter and set it in a drinking glass. Squeeze the juice from several lemons into the funnel filter. Remove the funnel filter and pour the juice into a shallow, wide dish or bowl. Let the juice stand for a day or two to evaporate some of the water and make a more concentrated solution.

In a similar way, squeeze the juice of a grapefruit through a funnel filter into a drinking glass. Then, pour into a shallow dish or bowl and allow the juice to evaporate for a day or two to make a more concentrated solution.

Make a cup of boiling water as you would when you make a cup of tea, using either a microwave oven or a tea kettle on a stove. Steep several tea bags in the hot water. Let cool. Then, pour the tea into a shallow dish or bowl and let some of the water evaporate for a day or two to make a concentrated solution.

Pour each of your three natural insecticide liquids in three spray bottles.

• *Testing effectiveness against fruit flies.* Fruit flies can often be seen hovering around a bowl of fruit. As fruit ripens, especially bananas, fruit flies can't resist them. Would spraying bananas with a natural substance such as grapefruit juice, lemon juice, or tea deter fruit flies?

Place several bananas in a bowl and allow them to ripen and draw fruit flies. When fruit flies appear, cut one of the ripened bananas lengthwise in half, and then cut each half lengthwise in half again, making four pieces of banana. Lay them out, side by side, and observe the behavior of fruit flies around the pieces.

Spray one piece of banana with your lemon concentrate, another with the grapefruit concentrate, and another with the tea concentrate. Leave the fourth piece alone as a control.

Lay the pieces out again, but leave four or five inches between each one. Observe the behavior of the fruit flies. Do any of the sprayed banana pieces appear to repel the flies?

• *Testing effectiveness against ants.* Locate an active anthill and capture several ants in a jar. On a board or outside on pavement, set down the jar. Remove the sprayer from the lemon-filled spray bottle and pour a six-inch diameter circle around the jar. Open the jar of ants. Do the ants cross the ring of lemon?

Repeat using the grapefruit juice, and then the tea concentrate. To ensure that the ants are not simply responding to a liquid, lay a line of plain water in front of the ants to compare their behavior to your insecticides. Observe the behavior of the ants.

Results

Write down the results of your experiment. Document all observations and data collected.

Conclusion

Come to a conclusion as to whether or not your hypothesis was correct.

Something More

Try your natural insecticides (concentrated lemon juice, grapefruit juice, and tea) with worms. Lay a line of your insecticide in front of a worm and observe its behavior. Does it cross the line? To ensure that the worm is not simply responding to a liquid, lay a line of plain water in front of the worm to compare its behavior to your insecticides. Are spiders affected by insecticides?

Project 32

The Nose Knows

Olfactory identification differences by age

Suggested Entry Category

• Behavioral & Social

Purpose or Problem

The purpose is to determine if the accumulation of life experiences by adults enables them to be better able to identify smells than young people.

Overview

Your nose receives a variety of smells each day, and many different smells throughout your lifetime. Your mind often associates a specific smell with a visual scene. The smell of seaweed may remind you of the sea, if you live by or have visited a seashore. The smell of pine may make you think of the Christmas holiday, when a live tree was decorated in your home. The aroma of baking cookies may remind you of being with your grandmother in her kitchen. The smell of a geranium may remind you of a plant in a neighbor's window sill.

These smells and their associations are imbedded in our minds as we go through life and enjoy many different experiences. Certain smells may not be identifiable to young people who have never experienced them. For example, the strong smell of mothballs is easily identifiable by many adults, but because much of our clothing today is made from synthetic materials, moth balls are not as popular as they once were. The smell of household lubricating oil may make an older adult male think of playing with an Erector construction toy set as a child. Burning leaves may remind an adult of diving into tall piles of raked leaves in the fall when they were little.

Smells play a very important part of our life experiences.

Hypothesis

Hypothesize that when presented with a group of common smells, a larger percentage of adults will be able to correctly identify the smells than will those of a younger age.

Materials' List

- Household oil
- Mothballs
- Leather product (such as a wallet)
- Cinnamon
- Clove
- Geranium plant
- Piece of cedar wood
- Pine tree branch
- Fresh-baked cookies
- Vanilla extract
- Ten adults to survey
- Ten young teenagers to survey
- Ten small jars with screw-on lids (such as baby food jars)
- Black construction paper
- Adhesive tape

Procedure

Find ten small jars with lids. Cover the sides of the jars with black construction paper, using adhesive tape, to prevent anyone from seeing inside the jars.

Gather ten items that have distinctive smells. Suggested items are given above in the Materials' List (mothballs, cinnamon, and so forth). Place a sample of each item in its own jar and screw on the lid.

Expose ten adults and ten young teenage students to each of the smells. Have each subject close their eyes when they smell the samples. Set up a log and keep a record of the number of items (and which items) each surveyed subject correctly identifies. You may want to later tabulate your results using a computer spreadsheet program with graphing capabilities.

Did a larger percentage of adults than students correctly identify the sample items?

Results

Write down the results of your experiment. Document all observations and data collected.

Conclusion

Come to a conclusion as to whether or not your hypothesis was correct.

Something More

1. Analyze the number of correct identifications of each item by adults, and then by students. Did most adults correctly guess particular items, while few students could? This may be a reflection on our changing society. Mothballs, as mentioned earlier, may not be familiar to many students, because moth balls are not as common as they once were in most households.

2. Compare your results by male and female subjects. Are certain items more easily identified by specific genders? For example, because more males probably play the sport of baseball than females, hypothesize that the smell of a new baseball glove may be more familiar to males than females.

3. Smells may cause memory recollection better than images.

4. The nose continues to grow throughout life and beyond!

Project 33

Germ Jungle

Checking for the presence
of bacteria on public surfaces

Suggested Entry Categories

- Biochemistry
- Microbiology
- Environmental Science

Purpose or Problem

The purpose of this project is to determine if bacteria exists on many commonly touched public surfaces. If so, then it is important that people are aware of the potential for bringing infectious bacteria into their homes after visiting a public place.

Overview

Think about the many surfaces your hands touch each day, at home, school, work, the library, and shopping malls. How many other people have touched those same surfaces? Money, shopping-cart handles, door knobs, telephones, ATM keypads, escalator handrails, gym equipment, and hundreds of other surfaces are touched by many different people throughout each day. A research study done by the University of Arizona examined

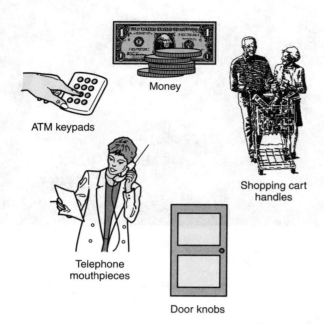

ATM keypads

Money

Shopping cart handles

Telephone mouthpieces

Door knobs

more than 800 public surfaces in three states, and found the potential risk of bringing harmful bacteria into the home is alarmingly high.

Bringing bacteria into your home can create a health risk for you and your family. Federal health officials estimate that 80,000 people die each year of infectious bacteria, making it the third leading cause of death in the United States (after cancer and heart-related problems).

Once bacteria enter your home, it can be spread around quickly. A person can contaminate and recontaminate frequently touched surfaces, such as refrigerator door handles, TV remote controls, and door knobs. That's why it is extremely important to wash your hands when you return home from a public place.

The results of this experiment should help make you, your family, and your friends more aware of the importance of washing your hands after visiting a public place, and

encourage you to use household disinfectant-cleaning products around kitchens and bathrooms.

Hypothesis

Hypothesize that the presence of bacteria will be found on many publicly touched surfaces, posing a potential health hazard to those who come in contact with those surfaces.

Materials' List

- Box of cotton swabs
- Eleven petri dishes
- Agar
- Masking tape
- Visit to a public shopping center or mall
- Pen and paper
- Possible adult supervision needed

Procedure

Collect 11 petri dishes and line the inside of them with agar. Your high-school science teacher may be able to supply you with these items. A *petri dish* is a shallow, round, transparent dish with an overlapping cover. *Agar* is a gelatin substance that is a food, or "culture," in which bacteria can be grown. Place a small piece of masking tape on the cover of each petri dish; the tape will be used as a label on which to write and identify the contents.

Open a new, sealed package of cotton swabs. As a control, wipe the swab in a petri dish with agar. On the masking tape on the cover, write, "cotton swap straight from the box." If no bacteria grows in this dish, we can safely assume the swabs are clean, and they are not introducing any bacteria into the other petri dishes.

Visit a busy shopping center. Identify ten highly touched surfaces, including the mouthpiece of a public telephone, door handles, ATM keypads, and other surfaces that you observe are being touched frequently.

For each item, use a cotton swab to wipe the surface. Then, wipe the swab into the agar. Dispose of the swab. Wrap a piece of masking tape around the petri dish, securing the lid to the bottom. On the dish's masking-tape label, write the name of the surface from which the sample was taken (ATM keypad, door handle, and so on).

When you arrive home, place all 11 petri dishes in a location that is out of the way, but that will remain at room temperature. The petri dish and agar make optimum growing conditions for bacteria. The agar gives bacteria food and proper moisture, and nothing will disturb or interfere with the bacteria's growth.

Make daily observations and keep a log for all the samples. Bacteria are too small to see, even with a standard microscope. But, if bacteria are present, you will eventually observe a large colony of growth on the agar.

It is important to realize that you are growing bacteria, which may pose a health hazard. Therefore, when this project is completed, keep the petri dishes sealed and dispose of them in the trash. Never open the petri dishes once they are sealed!

Results

Write down the results of your experiment. Document all observations and data collected.

Conclusion

Come to a conclusion as to whether or not your hypothesis was correct.

Something More

1. Check the effectiveness of the household disinfectant cleaners. Using cotton swabs, wipe different surfaces in your kitchen and bathroom areas, and then transfer the swabs to agar in petri dishes. Then, use a common household disinfectant cleaner to wipe the same surfaces. Again, use cotton swabs to wipe the surfaces, and transfer them to petri dishes. Compare the growth in the petri dishes.

2. If you live near a university, you may be able to make contact with someone who would allow you to use their electron microscope to identify the bacteria you have collected from various public surfaces. Unfortunately, bacteria are too small to be identified using an ordinary microscope.

Not 'til Christmas

Determining adherence to instructions by gender

Suggested Entry Category

- Behavioral & Social

Purpose or Problem

The purpose of this project is to determine if a behavioral difference exists between high-school male and female students regarding their attitude toward following instructions.

Overview

The Christmas holiday (or your birthday) is approaching, and a wrapped present with your name on it is sitting on the kitchen table. Can you resist shaking it and trying to determine its contents? Would you try to unwrap it just enough to peek inside, and then tape it back up again? Or, do you have the willpower and strong character to resist the temptation?

If a box labeled "Do Not Open" is placed in your school's cafeteria or another high-traffic area, do you think some students will open it? Do you think more males than females will open it? What do you think would motivate those students to open it: Curiosity? A lack of willpower? A tendency not to heed advice or to be defiant?

Hypothesis

Hypothesize that a larger percentage of high-school male students will attempt to open a box labeled "Do Not Open" than females (or hypothesize just the opposite).

Materials' List

- Plywood
- Framing pieces of 1×2 lumber
- Small wood screws
- Two small hinges
- Cabinet handle
- Paper and pencil
- Marker pens
- Adhesive tape (or glue)
- Camcorder (or video tape recorder with camera)
- Use of a saw (**have adult supervision if a power saw is used**)
- Permission to set up your project in the school cafeteria during lunch
- Possible adult supervision, if needed

Procedure

Using plywood for the sides and 1×2 or similar framing pieces of lumber, construct a box about one-foot square in size. The top side of the box must be hinged with a handle, so it can be opened. We want the box to look like it is very easy to open by simply lifting a handle.

With a large marking pen and paper (or use a computer and printer), print "Do Not Open" in big letters on the paper, and tape (or glue) it to the top of the box.

On another piece of paper, write "Please do not tell anyone what you saw inside this box" and place that paper inside the box at the bottom.

Get permission to set the box in a prominent place in the school cafeteria during lunch. The location should be where nearly everyone must pass by the box.

At a safe distance away, set up a camcorder or a camera with a video tape recorder, so you can record students without their knowledge as they are standing at the box. Turn the recorder on and let it run throughout the lunch period.

Later, use the video tape to total the number of males who approached the box and read the sign, and the number of females who approached and read the box. Also, log the number of males who opened the lid and the number of females who opened the lid.

Calculate the percentage of males who opened the lid. Divide the number of male students who lifted the lid by the total number of males who approached and read the sign. Then, multiply this answer by 100 (to arrive at the percent). Similarly, compute the percent for female students.

Something More

1. Survey the students who opened the box and ask if they can explain why they disobeyed the sign.

2. Hypothesize that adult teachers and staff will show greater regard for instructions and won't open the box, and they will have a lower percentage of "openers."

Results

Write down the results of your experiment. Document all observations and data collected.

Conclusion

Come to a conclusion as to whether or not your hypothesis was correct.

Space Farm

The effect of artificial gravity on radish-seed germination

Suggested Entry Categories

- Botany
- Earth & Space
- Environmental Science
- Physics

Purpose or Problem

The purpose of this project is to determine if roots and stems that grow out from germinating seeds are affected by an additional force perpendicular to the Earth's gravity.

Overview

Scientists are working on ways to grow food in space, as this will be very important if people are eventually to live on space stations for extended periods of time. One type of study that has been going on for years is *hydroponics*, which is growing plants with water, air, and nutrients, but no soil.

We have seen Hollywood science-fiction movies and TV shows depicting huge space stations that are rotating to make an artificial gravity inside. This is similar to the force that holds water in a pail when it is upside down as you swing it around with your arm or on a rope.

Normally, when a seed germinates, a stem grows skyward and a root extends down into the ground. Why? Is the root seeking water? Is gravity the factor that determines the direction of the sprouts? This may be important to know in the future as our quest for living in space becomes closer to reality.

Obviously, we will not be able to remove the effect of the Earth's gravity. Our experiment will determine if the spinning turntable exerts enough force in a different direction to act as an artificial gravity. *Geotropism* is the response of a plant to gravity.

Hypothesis

Hypothesize that seeds planted while being spun on a record player turntable at 33 RPMs will germinate, and their roots and stems will grow parallel with the Earth, rather than perpendicular as they normally would.

Materials' List

- Record player turntable
- One package of radish seeds
- Five small plastic drinking cups (about 5 ounces)
- Sheet of stiff cardboard or poster board
- Paper glue
- Potting soil
- Water
- Eyedropper
- Scissors
- 15 days of time

Procedure

LP (long-playing) record albums are 12 inches in diameter. Cut a 12-inch diameter circle from a piece of stiff cardboard or a sheet of poster board. With a pencil point or scissors, poke a hole in the exact center of the board, so it will slide down over a turntable's spindle.

Pour potting soil into five small plastic drinking cups, to a depth of one inch. Moisten the soil with water, but do not flood it.

Push one radish seed into the soil in each of four plastic cups. In the fifth cup, push three or four seeds. The fifth cup is our control cup, which will prove that the seeds in the pack and the growing environment are viable. Set this fifth cup aside in an out-of-the-way area in the room where your project will be placed.

Push the cardboard disc over the spindle and down onto the platter of a record player turntable. Be sure the disc moves when the turntable is spinning, and that the disc's hole is not too tight around the spindle.

At four opposite "sides" of the cardboard circle, place a few drops of glue (at 12, 3, 6, and 9 o'clock positions). Set a cup on each spot of glue. Let dry several hours.

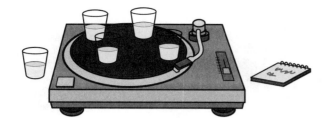

Power up the turntable, setting it at a speed of 33 RPMs (revolutions per minute). Daily, add water equally to all five cups to keep the soil slightly moist. You may want to use an eyedropper to ensure an equal quantity of water is added to all cups. The turntable must run continually, except when you're adding water or making observations.

Make observations daily for 15 days, and write down your observations. Did the seeds germinate? If so, which direction (relative to the Earth's surface) did the roots grow? Which direction did the stems grow?

Results

Write down the results of your experiment. Document all observations and data collected.

Conclusion

Come to a conclusion as to whether or not your hypothesis was correct.

Something More

1. If the seeds germinated, did the stems grow toward the turntable's spindle and the roots away from it?

2. Set the turntable to 45 RPMs and repeat the experiment. The higher turntable speed will exert more force on the seeds. Compare the results of seed germination at 33 RPMs and 45 RPMs.

Cooled Off

Comparison study between the cooling effect of evaporating water and alcohol

Suggested Entry Categories

- Engineering
- Physics
- Medicine & Health

Purpose or Problem

The purpose is to determine if the temperature of an object can be cooled more by evaporating alcohol surrounding it than by water or air.

Overview

The process of evaporation produces a cooling effect. Pour a few drops of water onto the back of your hand. Next, swing your arm back and forth. Do you feel your skin getting cooler?

Now pour a few drops of alcohol onto the back of your hand and swing your arm again. Does your skin feel even cooler?

Alcohol evaporates faster than water and, thus, creates a cooler temperature. When children run dangerously high fevers, doctors sometimes give them a bath in alcohol to reduce their external temperature, which, in turn, reduces their internal temperature.

How much change in temperature occurs between alcohol and water evaporation?

Hypothesis

Hypothesize that a greater percent change in temperature will take place when alcohol is evaporated rapidly than when water is evaporated.

Materials' List

- Two-foot-square piece of plywood
- Several pieces of 1×2 lumber
- Wood screws
- Saw
- Screwdriver
- Three thermometers
- Duct tape or adhesive tape
- Six thumbtacks
- Old T-shirt or flannel shirt
- Alcohol
- Water
- Electric fan

Procedure

Mount three thermometers side-by-side onto a piece of two-foot-square plywood, using adhesive tape or duct tape. Do not cover either the thermometer's bulb or the area on the scale around the 70°F mark. The board needs to stand vertically, so using a few pieces of small 1×1 or 1×2 lumber and wood screws, construct support struts onto the plywood to enable it to stand upright.

Cut three identical four-inch squares in a piece of an old T-shirt or flannel shirt. Fold each in half, and then fold in half again. Slip a piece of cloth over the bulb of each thermometer. Use thumbtacks to hold the cloths in place, pushing one thumbtack on either side of each bulb through the cloth and into the board.

Set the thermometers on a table. Place an electric fan in front of the thermometers, positioning the fan so a high volume of moving air hits the bulbs.

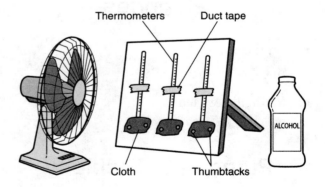

Thermometers   Duct tape

Cloth   Thumbtacks

Let the thermometers sit for ten minutes to stabilize at room temperature. Label each thermometer A, B, and C. Write down the temperature reading on each thermometer. It does not matter if they are not precisely calibrated, and they do not all need to have exactly the same temperature.

Soak the cloth around thermometer A's bulb with rubbing alcohol. Soak the cloth around thermometer B's bulb with water. Leave C dry.

Turn on the fan. Monitor the three temperatures for several minutes. Record the lowest temperature reading on each thermometer.

Calculate the percent change in temperature of each thermometer. Percent change is calculated by subtracting the lowest temperature reading from the room-temperature reading, dividing that answer by the original room temperature reading, and then multiplying by 100 (to put the answer in percent). For example, if a thermometer reads 72°F at room temperature, and then drops to 68°F during the experiment:

$$72°F - 68°F = 4°F$$

$$(4/72) \times 100 = 5.56\%$$

Did the alcohol-soaked bulb drop the lowest? Why doesn't it matter whether all the thermometers were calibrated to read exactly the same at room temperature?

Results

Write down the results of your experiment. Document all observations and data collected.

Conclusion

Come to a conclusion as to whether or not your hypothesis was correct.

Something More

1. Does the velocity of the moving air affect the temperature of evaporation? Repeat the experiment, once with the fan on its lowest setting, and then once on its highest setting. Graph your results, drawing them on graph paper or using a computer spreadsheet program that prints graphs from entered data.

2. In an environment where humidity is high, will the evaporation of alcohol and water still result in a cooling effect? Repeat the experiment in a steamy environment.

3. How does the fact that evaporation causes a cooling effect relate to wind-chill factor?

4. How do air conditioners work?

Pass the Mold

A study on the capability of common bread mold to be transferred from one food to another

Suggested Entry Categories

- Environmental Science
- Microbiology
- Health & Medicine

Purpose or Problem

The purpose is to study the capability of common bread mold to be transferred through the air and contaminate other foods.

Overview

Have you ever opened the refrigerator to find an old, leftover food that had mold growing on it? Have you ever opened a loaf of bread that has been in the bread box for a long time and seen green mold spreading over it?

Mold growth is quite common, but many people are very allergic to mold. It may aggravate allergies or respiratory problems.

Can bread mold become airborne and, therefore, become a health hazard?

Hypothesis

Hypothesize that common bread mold can be transferred by air to contaminate other foods. (You may also choose to hypothesize that mold does not become airborne.)

Materials' List

- Fresh bakery bread
- Orange
- Banana
- Sealable plastic food bags
- Eyedropper
- Water

Procedure

Grow bread mold on three slices of bread. You will need bread that does not contain "mold inhibitors," which are ingredients in most store-bought bread. Try bread from a local bakery and ask if it contains mold inhibitors, or try a piece of rye bread, which molds more easily.

Using an eyedropper, place 15 drops of water on each slice of bread. Place the bread in sealable plastic-food wrap bags.

Let the bags sit for several days until a large amount of mold grows on them. Once a large amount of mold has grown on the three slices of bread, proceed. **If you are sensitive to mold or have allergies, wear a mask when you do the next step.** Paper filter masks are available at the hardware store or your local pharmacy.

Take two other slices of fresh bread. Add 15 drops of water to each slice. Place one in a sealable plastic bag. This will be the control slice.

Remove one of the slices of moldy bread from its bag. Hold it over the other slice of fresh bread and shake it vigorously. Place the fresh bread in a sealable plastic bag as a control. Put the mold-covered bread back into the bag, seal it, and discard it in the trash.

Cut an orange in half. Place one-half in a sealable plastic bag as a control. Shake a piece of moldy bread over the other half of the orange. Place the moldy bread back in its bag and discard it in the trash. Place the orange half in a plastic bag.

Peel a banana, and cut it in half lengthwise. Place one half in a sealable plastic bag as a control. Shake a piece of moldy bread over the other half. Place the moldy bread back in its bag and discard it in the trash.

Monitor the slices of bread, orange, and banana for several days. Does mold form on them? If so, does it form faster or in more abundance on the food over which the moldy slices of bread were shaken?

Results

Write down the results of your experiment. Document all observations and data collected.

Conclusion

Come to a conclusion as to whether or not your hypothesis was correct.

Something More

Try shaking a piece of moldy bread over other food substances, such as flour and corn starch. Add moisture and place in sealable containers. Does mold grow?

Hardwood Café

Determining if bracket fungi are parasites or saprophytes

Suggested Entry Categories

- Botany
- Environmental Science
- Microbiology

Purpose or Problem

The purpose is to discover whether bracket fungi are parasites or saprophytes.

Overview

"Bracket" or "shelf" fungi can be found in wooded areas growing on the sides of trees, fashioning themselves as little shelves, perhaps for elves! Fungi do not photosynthesize, as do other plants. They get their nourishment from a host they live on. If a plant gets its nourishment from a host organism that is dead and decaying, it is called a *saprophyte*. If the host is a living organism, the feeding plant is called a *parasite*.

Are bracket fungi saprophytes, parasites, or both?

Trees have tiny tubes that transport water, nutrients, and waste throughout their system. These tubes are called *xylem* (which transport wastes) and *phloem* (which transport food). Trees grow from the outer layer just beneath the bark. The bark is not living. Our project is to locate bracket fungi and carefully chip away at the bark of the host tree and see if any "roots" or threadlike structures penetrate through the bark and into the live layer of the tree. If this is the case, then bracket fungi is most likely a parasite. If not, it is most likely a saprophyte.

Hypothesis

Hypothesize that bracket fungi are saprophytes (or hypothesize that they are parasites, or that they are found on both live and dead trees).

Materials' List

- Wooded area
- Chisel

Procedure

In a forest or wooded area, locate trees on which bracket fungi are growing. Using a chisel, carefully pry pieces of bark off the tree around the bracket fungi. Try to determine if any part of the fungi extends through the bark and into the soft, live layer of the tree. To avoid injuring the tree, do not remove too much bark.

Results

Write down the results of your experiment. Document all observations and data collected.

Conclusion

Come to a conclusion as to whether or not your hypothesis was correct.

Something More

Carefully search through a large area in the forest, noting any presence of bracket fungi. Are the trees where you find the fungi dead or alive, or do you find them on both dead and live trees?

Web Crawlers

Determining the effectiveness of various Internet search engines

Suggested Entry Categories

- Behavioral & Social
- Engineering (Software/Communications Engineering)
- Math & Computers

Purpose or Problem

The problem is finding specific information among the vast data available on the Internet.

Overview

The invention of the Internet has become one of the most life-changing and society-changing tools of our lifetime. Anyone who needs to do research or who requires information on a particular subject has access to a wealth of data and up-to-the-minute news and published research on that subject.

The problem is this: Since the Internet is so vast, how do you search through all the information to uncover the data that is of interest to you?

An Internet *search engine* is a software program used to look constantly through the entire World Wide Web (the Web) and it indexes all the sites found there. The engine utilizes indexing software, sometimes referred to as *robots*, *bots*, or *spiders*. These spiders and bots "crawl" around the Web looking for new or updated pages. They travel from address to address, known as Uniform Resource Locators (URLs), until they have visited every site on the Web.

This is a monumental task, though, even for computer software working at fantastically high speeds. Because of the huge number of sites, it may take a long time for spiders and bots to get to every site. Therefore, it is possible for one search engine to give you different results than another.

Hypothesis

Hypothesize that when searching for a particular subject on the Internet, different search engines will yield different results. Therefore, to thoroughly scan the Internet database for specific information, you should always use more than one search engine.

Materials' List

- Computer connected to the Internet

Procedure

Make a list of a dozen specific subjects (perhaps a certain type of bacteria or a type of unusual plant). With a computer connected to the Internet, use various search engines to search for that topic. Write down how many references each engine found. Then, look at each site that is referenced, and count how many are relevant and how many do not apply to your topic.

For example, a popular term for one stock market strategy is "rolling stock." *Rolling stock* identifies a stock whose price is fluctuating up and down in a narrow range, enabling a stock market trader to buy the stock when it is at the low part of its range, and then sell it when it rolls up to a resistance price. However, if you use search engines to look for rolling stock, the results will also include links to the railroad industry, because that term is used to mean transporting cattle by rail.

Popular search engines include Google, Ask Jeeves, Excite, Infoseek, Lycos, Alta Vista, Webcrawler, and Yahoo!.

Did different search engines give different results? Which search engine produced the most consistently relevant results?

Results

Write down the results of your experiment. Document all observations and data collected.

Conclusion

Come to a conclusion as to whether or not your hypothesis was correct.

Something More

Technically, some of the software we use to search the Web is not truly a search engine, but rather a "directory." A true search engine only needs the address of a web site. Then, an indexing agent (like a spider) does the rest. A directory requires the owner of a web site to provide the directory with a list of categories under which the site should be catalogued. Excite, Infoseek, Lycos, Alta Vista, and Webcrawler are example of search engines. Yahoo! is an example of a directory, just as The Yellow Pages in your telephone book is an example of a directory. Compare the results of searching for a subject using a search engine and a directory.

Project 40

Night Watch

Circadian rhythms: Training a house plant to be awake at night

Suggested Entry Categories

- Botany
- Environmental Science

Purpose or Problem

The purpose is to see if a plant's natural biological rhythms are upset (as evidenced by an observable change in the plant) when normal daylight and dark nighttime periods are reversed.

Overview

Are you tired at 3 A.M., and wide awake with lots of energy at 3 P.M.? Our bodies have a natural pattern of sleep and waking times. Have you ever tried to switch your sleep/wake time, so you are awake in the night and sleep in the day? Certain jobs, such as factory or police swing-shift workers, have alternating day/night work schedules. People who travel to foreign countries often suffer jet lag, as their bodies' "circadian rhythms" are upset by the different times of sunlight and eating meals. Jet lag can cause many problems, including fatigue, cloudy thinking, and irritability.

Plants, animals, and people have daily biological rhythms. As far back as the fourth century BC, the scribe of Alexander the Great, Androsthenes, observed that the leaves of certain trees opened during the day and closed at night. Leaves of the heliotrope plant have a similar action, as do day lilies. Bees visit flowers at specific times of the day. Rhythms in other animals are also now well known.

Just as people need a period of rest at some time during a day/night cycle, so do plants. If we change the time of day a plant is exposed to light, will that result in a noticeable effect on its growth or health?

Hypothesis

Hypothesize that changing the time of light and dark during a 24-hour period will cause an observable effect on a plant within an eight-week period. Observable effects include growth, color, size, turgor, and general healthy appearance. (You may want to hypothesize that no noticeable effects will occur.)

Materials' List

- Nine equally healthy house plants
- Two dark closets
- Two plant grow light bulbs
- Kitchen measuring cup
- Eight weeks of time

Procedure

Obtain nine equally healthy, identical types of house plants. Obtain two equal wattage plant grow lights, available at your local nursery or garden center. These indoor lights supply the proper wavelengths of light to grow plants without sunlight. Locate two closets that are on the same floor in your home. They must have doors on them, so when the doors are closed, the closets are completely dark.

Place three plants in each closet. Place a plant grow light in each closet. Set the three other plants in an out-of-the-way area in a living room or dining room, where they will receive ample light, but not be in direct sunlight from a window. The plants in the living room will receive normal light during daylight hours and normal darkness in the room at night.

For the plants in one closet, turn the plant light on from 8 A.M. to 8 P.M. daily, leaving them in the dark during the evening.

In the other closet, turn the plant light on from 8 P.M. to 8 A.M., so the plants are in the dark during the day, but receive light during the night.

Monitor the plants for eight weeks. When you water the plants, use a kitchen measuring cup, so you give an equal amount of water to each of the nine plants. Keep the soil moist, but not drenched. Another constant—temperature—is assumed to be about equal in the closets and living room.

Results

Write down the results of your experiment. Document all observations and data collected.

Conclusion

Come to a conclusion as to whether or not your hypothesis was correct.

Something More

1. Deprive a plant of a rest period. Compare a plant that gets a normal amount of light during the day and dark at night to a plant that receives constant light and is never allowed a rest cycle. Observe the appearances of the plants after eight weeks under these conditions. Use at least four plants, two under each condition.

2. Experiment with plants that demonstrate a very visible biological rhythm (for example, day lilies and morning glories).

Time for the Concert

A study of the effect of temperature on the chirping of crickets

Suggested Entry Categories

- Behavioral & Social
- Environmental
- Zoology

Purpose or Problem

The purpose is to see if temperature affects the chirping activity of crickets.

Overview

On warm nights in the spring or summer, in quiet wooded areas, the sound of chirping crickets fills the air. Do the number of chirps increase as the temperature increases? Do you recall ever hearing crickets on cold evenings? Is there a cool temperature below which crickets will not chirp? If so, you can get an idea of the outdoor temperature by listening to the sound (or the absence of sound) of chirping crickets.

Hypothesis

Hypothesize that temperature has an effect on the chirping sound made by crickets.

Materials' List

- Several common house crickets
- Pet "critter cage," empty aquarium tank, or similar container
- Pet heat rock
- Thermometer
- Ice cubes
- Sealable plastic food bags
- Tape recorder
- Paper and pencil

Procedure

Obtain several common household crickets. They can usually be purchased at local pet shops, because they are used as *feeders* (food for larger pets). Place the crickets in a container with solid, clear sides. "Critter cages," used for snails, or empty aquarium tanks have the kind of sides you want. You may want to make your crickets more comfortable by introducing food and a water dish into the container. Check with your pet shop personnel for assistance.

Place a thermometer in the container with the crickets. Record the current temperature and make note as to whether or not the crickets are chirping.

Raise the temperature by placing a heat rock in the container. Heat rocks are available at pet shops, and they are used for many reptiles, such as lizards and iguanas. Record the temperature as it increases. Use a tape recorder to record the chirping sounds. Talk into the microphone, announcing the temperature as each degree changes.

Next, lower the temperature in the container by removing the heat rock and placing sealable plastic food bags filled with ice cubes into the container. Continue to monitor the temperature and record the sounds.

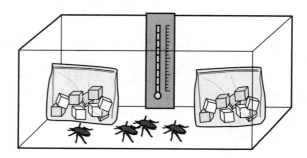

If your crickets do not chirp, could it be they are nocturnal? Are the crickets waiting for evening or for the sky to get dark?

Is there a temperature below which the crickets are silent? Do crickets chirp louder or with an increased number of chirps as the temperature increases? Use the tape recorder to compare sounds made at different temperatures.

Results

Write down the results of your experiment. Document all observations and data collected.

Conclusion

Come to a conclusion as to whether or not your hypothesis was correct.

Something More

Grasshoppers are said to chirp loudest at 95°F and are unable to chirp when the temperature falls below 62°F, unable to fly below 45°F, and unable to jump below 36°F. Conduct an experiment to see if these statements can be supported by your experimental data.

Project 42

Flying, Walking, Crawling

Natural bait to keep pests at bay during picnics

Suggested Entry Categories

- Behavioral & Social
- Zoology

Purpose or Problem

The problem is that people who enjoy outdoor picnics are often plagued by insect pests, including flying insects, as well as ants and other crawling creatures. The purpose of this project is to do a study on popular brands of soft drinks regarding their attractiveness to picnic pests.

Overview

Along with the great outdoors comes a great abundance of picnic pests, that is, insects and bugs, that can take some of the enjoyment out of a day by the pool or a backyard picnic.

We noticed that when we throw empty soda (soft drink) cans in our recycle trash can (which does not have a lid), bee-like flying insects appear, flying around the cans. They are not present when other types of jars and cans are discarded. Is something in soda that is attractive to these insects? Could soda also be attractive to other types of pests? If so, soda could be poured out in pie pans and placed at various points around the perimeter of a backyard, hopefully, to act as bait and keep pests away from people in the yard.

Hypothesis

Hypothesize that one or more brands of soda will be attractive to flying-insect pests. You may also hypothesize that whatever brand of soda is attractive to one type of insect will also be attractive to other types of insects.

Materials' List

- Warm day
- Seven different types of soda
- Seven pie pans
- Use of a backyard
- Picnic table

Procedure

On a warm day, when mosquitoes or other flying insects are present in a backyard, set seven pie pans on a picnic table. Disposable aluminum pie pans can be used, but if they have holes in the bottom, line them with tinfoil to make them watertight.

Purchase a selection of seven different types of soda. Include a mixture of dark colored, clear, diet, and different flavors

(cola, root beer, orange). Pour one type of soda into each pie pan.

Monitor the pans for several hours. Observe any insects that appear to be "hanging around" the pans. Make notes on which pans are visited the most. See if any insects are floating in the soda.

If few insects visit your pans, repeat the experiment in the early evening before the Sun sets. The cooler temperature of evening may bring more insects out of hiding.

Did any of the soda pans attract visitors? Did one type attract more than another? If so, what do you think caused the attraction (color, flavor, type of sugar—diet/regular—and so forth)?

Pick up a field guide on local insects at a book shop or the library, and try to identify any insects you see.

Results

Write down the results of your experiment. Document all observations and data collected.

Conclusion

Come to a conclusion as to whether or not your hypothesis was correct.

Something More

1. If the soda-filled pie pans are placed on the ground, will they also act as bait to deter crawling pests (ants, crickets, and so forth) from a picnic blanket you place on the ground?

2. Repeat the project in the late evening, when nocturnal pests are active. Are more pests found around the soda pans at night than during the day? Are they different types of insects?

Project 43

High-Tech Times

A study of the willingness of people in different
age groups to adapt to new technology

Suggested Entry Categories

- Behavioral & Social
- Math & Computers

Purpose or Problem

New technology is invading our lives at an
ever-increasing rate. While many people find
new technology exciting and can't wait to get
involved in it, some people may feel intimidated
by it, or feel that it only complicates their lives
and puts more stress on them.

Overview

The technological advances we have seen,
especially in the last few decades, have
changed the way we do many things. The
Internet as a source for news, entertainment,
research, buying goods, and stock market
trading has revolutionized the home
computer. The computer itself has changed
the way many of us work and play. VCRs,
cellular phones, ATM bank machines, e-mail,
and a host of other new inventions have made
our work and pleasure time easier and more
enjoyable.

Hypothesis

Hypothesize that when people of different age groups are surveyed about their use of new technology, a greater percentage of younger people will answer "Yes" to their use of it than will older people.

Materials' List

- Paper and pencil
- 20 people between the ages of 16 and 30
- 20 people between the ages of 31 and 49
- 20 people over age 50

Procedure

Create a short list of questions to ask in a survey about the use of new technology. The survey should be similar to the following example.

Answer Yes or No to these questions:

1. Do you own a personal computer?
2. Do you use e-mail regularly?
3. Do you bank online?
4. Do you know how to program and set the time on your VCR?
5. Do you have a cellular phone?
6. Do you use a bank ATM machine?

But, maybe not everyone believes these inventions are so wonderful. Do you think older people may be more resistant to technology, wanting to do things the way they have always done them? Perhaps they like doing things the way they are comfortable with, or perhaps they are confused by new technology.

Many inventions have taken place in the last few years that we take for granted. For example, if you watch reruns of the original *Star Trek* episodes on TV, you are familiar with the automatic opening of doors as people approached them. Today, we don't give a second thought to doors opening by themselves as we walk into a department store or a supermarket!

This project will conduct a survey of different age groups and ask them about their use of new technology. This study will give insight into how people of different ages are coping with these changes in society.

If you can't get 20 people in each group to participate in the survey, get as many as you can. Calculate the percentage of people who answered Yes to each question in each group. For example, suppose 15 out of 20 people surveyed in the first group answered Yes. Divide the number of Yes answers by the total number of people, and then multiply by 100 to get the percent:

$$(15/20) \times 100 = 75\%$$

Results

Write down the results of your experiment. Document all observations and data collected.

Conclusion

Come to a conclusion as to whether or not your hypothesis was correct.

Something More

Were there any questions in which a greater number of older people had a higher percentage of Yes answers?

Commercial TV

A comparison of programming to advertising content

Suggested Entry Categories

- Behaviorial & Social
- Math & Computers

Purpose or Problem

The purpose is to determine if there is more advertising content per hour during the day than during an hour of evening prime-time viewing (8 P.M. to 11 P.M.).

Overview

Advertising is important. Commercials inform us of sales and new products that may be of help or enjoyment to us. Ads can also let us know how to save money. Nevertheless, sometimes TV advertising seems to take up a large portion of the hour compared to program content time. On average, is the percentage of commercial-to-program content higher during the day than compared to the night? And is that average consistent among the various channels?

Hypothesis

Hypothesize that more TV commercials are run per hour in the daytime than in the evening.

Materials' List

- Stop watch or a watch with a second hand
- VCR with blank video tape
- TV with cable-channel accessibility

Procedure

Collect data on the amount of commercial (advertising) time during a typical hour of daytime programming (2 P.M. to 3 P.M., for example) and a typical hour of prime time (8 P.M. to 9 P.M., for example). Collect this data over a five-day period, Monday through Friday. Add the data and divide by five to find a daily average.

If you are not home during the day, use a VCR with a timer set to record from 2 P.M. to 3 P.M. Then, when you get home in the evening, you can watch the program and collect your data.

Because some channels are only seen over cable, and some channels are also broadcast over the airways, data must be collected on both and averaged together. Collect data (day and night) for a cable channel (Discovery Channel, the Learning Channel, CNN Headline News, and so forth), and one for an over-the-air station (ABC, NBC, CBS). Average the results together.

Graph your results for presentation, using a computer program.

Results

Write down the results of your experiment. Document all observations and data collected.

Conclusion

Come to a conclusion as to whether or not your hypothesis was correct.

Something More

1. What is the average length of time of a commercial spot (15, 30, or 60 seconds)? Compare the ad-time lengths during the day, the evening, and between cable channels and the big over-the-air networks (ABC, NBC, and CBS).

2. Can you think of ways to use the time during commercial breaks, other than going to the kitchen for a snack? A gallon water jug could be filled (or partially filled to weigh less) to be used during commercials as a dumbbell for getting mild exercise during commercials.

3. The data you collected is baseline data, which someone could use five or ten years from now in a comparison. Can you find out if someone gathered this data ten years ago? If so, compare it to your data.

Project 45

Sold on Solar

The temperature in a climate as it relates to the amount of possible usable sunlight

Suggested Entry Categories

- Earth & Space
- Physics

Purpose or Problem

The purpose is to prove that temperature is not a factor in determining if a location is acceptable for solar-energy applications.

Overview

Making use of energy from the Sun has many important benefits. *Solar energy* is a clean, renewable, and free source of fuel. It can be used to generate electricity and make hot water for our homes. *Solar panels* can be mounted on the roof of a home to assist in supplying hot water for the family, thus reducing the home's hot water utility bill.

Do solar panels only work for homes located in warm or hot climates? No. Temperature is not a factor, as this project will prove. If you live in a part of the country where it is cold in the winter, think about sitting at a table next to a bright, sunny window. You can feel heat on your body. Think about getting into a car on a sunny, cool day. The car's interior is warmer than the weather outside. Think about how cold it is in the Arctic, yet it is a very sunny region.

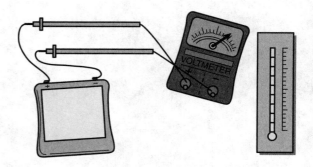

Hypothesis

Hypothesize that temperature is not a factor in the amount of usable sunlight that can be gathered in an area needed for solar-energy applications.

Materials' List

- Photovoltaic cell
- DC voltmeters with a millivolt scale
- Thermometer
- A day when the temperature is much colder outdoors than inside
- Daily newspaper or almanac with the times of sunrise and sunset
- Pencil

Procedure

Connect a DC (direct current) voltmeter with a millivolt scale across the leads (wires) extending from a solar cell. Small solar cells are available at local hobby shops and electronic parts chains. Be aware that the output of the solar cell has a polarity; it has

a plus (+) and a minus (–) terminal. Be sure to wrap the bare end of the plus lead of the solar cell to the plus test probe on the voltmeter.

On a cold day, set the solar cell and voltmeter outside in an unobstructed area that is exposed to direct sunlight. Set a thermometer next to it. Wait several minutes for the thermometer to adjust to the temperature. Record the voltage indicated on the voltmeter and the temperature on the thermometer.

Move the solar cell, voltmeter, and thermometer indoors, and place them in a sunny window. Wait a few minutes to allow the thermometer to adjust to the indoor temperature. Record the voltage and thermometer readings. What is the difference in temperature? While there was a big difference in temperature, was there a significant difference in voltage?

Another factor in determining if a location is favorable for the installation of solar panels is the amount of possible daily sunshine. Use the formula:

$$\text{percent of possible sunshine} = \frac{\text{minutes of the day the Sun could cast a shadow}}{\text{total minutes of actual daylight}}$$

The "total minutes of actual daylight" can be found in your daily newspaper by finding the time of sunrise and sunset, and then calculating the time difference between them and converting that time to all minutes. The "minutes of the day the Sun could cast a shadow" can be determined by taking a pencil or another object outside just after sunrise and holding it a few inches above the ground. When a clear and discernible shadow can be seen, record the time. Similarly, as sunset approaches, record the time when a shadow is no longer discernible. Subtract the two times to find the minutes of strong sunshine. Then, compute the percent of possible sunshine by dividing shadow time by total daylight time.

Results

Write down the results of your experiment. Document all observations and data collected.

Conclusion

Come to a conclusion as to whether or not your hypothesis was correct.

Something More

1. If you have a relative who lives a far distance (in another state) from you, ask them if they would do the same experiment as you just did. Compare their results to yours.

2. Does the glass in a window block some frequencies in the spectrum of light? Does it block or pass ultraviolet light? Does it block or pass infrared light?

3. Is the area you live in suitable (economical) for hot-water solar collectors or photovoltaic solar-cell applications? What percentage of days per year does your location experience overcast skies? Gather data from weather sources using the Internet, National Oceanic and Atmospheric Administration (NOAA), or local weather authorities. How would you construct a device that would monitor the amount of cloudiness that occurs during a day?

Getting to the Root of the Problem

A study of the effect of low water on radish seedling root systems

Suggested Entry Category

- Botany

Purpose or Problem

The purpose is to discover if low-moisture levels will cause the roots of radish seedlings to grow more abundantly than those grown in soil with high-moisture levels. A secondary purpose is to discover the optimum amount of water for radish seedlings.

Overview

Plants gather water and nutrients from soil through their root systems. Roots extend downward and outward to gather water. What if water is scarce? Will a plant's roots grow longer and will there be more of them, as they search for water?

Hypothesis

Hypothesize that the root system of radish seedlings will grow more abundantly (more mass and/or longer in length) in soil where moisture is low compared to soil where moisture is high. Also hypothesize that a point exists beyond which any further reduction in moisture will not cause more root growth but, instead, will cause health damage to the plant.

Materials' List

- Paper towels
- Radish seeds (50 of them)
- Ten 8-ounce plastic drinking cups
- Potting soil
- Kitchen teaspoon measure
- Marker or pen and masking tape
- Water
- Several weeks of time

Procedure

Germinate 50 radish seeds by covering them between layers of paper towel and keeping them warm and moist for several days.

Once the seeds have germinated and begin to sprout roots and stems, select 30 of the best seedlings and discard the rest.

Fill ten 8-ounce plastic drinking cups ¾ full of potting soil. Plant three seedlings in each cup.

Using a marker or a pen with masking tape, label each cup with a letter of the alphabet: Cup A, Cup B, Cup C, and so on through Cup J.

Group the cups: Cups A, B, and C make up Group #1. Cups D, E, and F make up Group #2. Cups G, H, and I make up Group #3. Cup J will stand alone.

Group #1 will receive water daily. Cup A will receive one teaspoon of water. Cup B will get two teaspoons. Cup C will get three teaspoons.

Group #2 will receive water every other day. Cup D will receive one teaspoon of water. Cup E, two teaspoons. Cup F, three teaspoons.

Group #3 will only receive water every third day. Cup G, one teaspoon. Cup H, two teaspoons. Cup I, three teaspoons.

The seedlings in Cup J will be kept soaked with water daily, adding enough water to keep the soil thoroughly wet.

We are using three seedlings per cup to increase our sample size, thus giving more credence to our results.

Set the cups in an area of equal lighting and equal temperature (room temperature).

After many weeks, remove the seedlings from their cups and rinse the soil off their roots. Compare the root systems of all the plants.

Was your hypothesis correct? Was there a point beyond which any less amount of water resulted in poor growth or healthy appearance of the plants? How was the health of the seedlings in Cup J? Is it possible to overwater a plant?

Results

Write down the results of your experiment. Document all observations and data collected.

Conclusion

Come to a conclusion as to whether or not your hypothesis was correct.

Something More

If a house plant is consistently given water only down one side of the flower pot, will the roots eventually grow toward that side (hydrotropism)?

Index